THE BEHAVIOR OF THE EARTH

The
BEHAVIOR
of the EARTH

Continental and Seafloor Mobility

CLAUDE ALLÈGRE

Translated by
DEBORAH KURMES VAN DAM

HARVARD UNIVERSITY PRESS
Cambridge, Massachusetts, and London, England   1988

This book is printed on acid-free paper, and its binding
materials have been chosen for strength and durability.

Library of Congress Cataloging in Publication Data

Allègre, Claude J.
  The behavior of the earth.

  Translation of: L'écume de la terre.
  Includes index.
  1. Plate tectonics. I. Title.
QE511.4.A4513 1988   551.1'36   87-31102
ISBN 0-674-06457-7 (alk. paper)

CONTENTS

PREFACE

SCIENTIFIC theories are like talented artists: once recognized their merits seem so obvious that their success is assumed to be due only to their excellence. In science especially, new ideas are seen as an inevitable and unshadowed enlightenment, and the fact that the process of discovering them was slow and chaotic is forgotten. The newer an idea, the more shocking it is, and the more it disturbs those who established their reputation before its emergence, as well as those whose intellectual security is upset by it. Originality is a highly esteemed virtue as long as it is not *too* original. Any innovation that goes beyond a certain threshold tends to be pushed to the sidelines, even discounted. To paraphrase René Girard's notion of imitative behavior, the only tolerable originalities are those that are differentially original (in a mathematical sense). The quantum jump in the evolution of scientific ideas is taboo because it violates the principle of *mimesis.*

The idea of continental drift was just such a quantum jump. Although it was not new when it took root in the earth sciences in 1961, it set off a fierce and passionate battle, in many ways comparable to the one that accompanied the emergence of evolutionist ideas in biology. As Darwinian theory is central to all of modern biology, so the theory of continental mobility is the pivot of the revolution the earth sciences have experienced in the past three decades. Without it modern geology would not be conceivable today. Few disciplines have in so short a time undergone such a radical transformation in mode of thought, concept, or method of approaching data.

Continental drift entails the idea of the re-creation of the ocean floor, and therefore of dynamism in the mantle and a coupling between the interior and the surface of the earth. In light of this theory the earth becomes a living, changing entity whose "physiology" can be understood only by studying it globally. Traditional geology consisted mainly of rock classifications, the timetable of geologic eras, cross sections and maps. Its vocabulary was as barbarous as it was esoteric. Today's geology is much more vivid and less static.

For a long while it was thought that movements of the earth's crust were essentially vertical, causing a trench to open here, a mountain to rise there. The only lateral displacements taken into account were those of the sea. Since the birth of geology it has been recognized that during certain epochs the sea covered vast areas of the continents, depositing sedimentary strata containing fossils. For example, the sea invaded the Paris and Aquitaine basins 200 million years ago (in the Triassic period) and did not retreat until a few dozen million years ago (at the end of the Tertiary). On an even more grandiose scale the sea invaded the entire African shield 500 to 150 million years ago. Only a few archipelagoes in the Ahaggar and in the south of Morocco stuck out from the vast expanse of water. Then, under the influence of factors that even today are not well understood, the sea retreated, allowing the continents to emerge. Marine transgressions and regressions were sometimes local, affecting only a small fraction of the land area, and sometimes general, affecting most of the continents.

The length of time the sea remained in a given region is indicated by the depth of sediments deposited one on top of another. The presence of animal and vegetable remains in them makes it possible to specify the age of a stratum and the ecological conditions that prevailed at the time of its deposition. Sedimentary layering thus constitutes the "text" from which a person who understands the language can decipher geologic history.

Because marine transgression and therefore sedimentation was episodic, the record on every continent is fragmentary and scattered. Traditional geology attempted to reassemble these archives, to match them up with one another and to read the messages they contain. This difficult, often tedious work made it possible for geologists to reconstruct the successive geographies of the earth. The continental massifs—blocks of hard, ancient rocks as are found in the north of Canada, a good part of Africa, the center of Australia, Scandinavia, and Central Asia—were assumed to be fixed. Around them the advances and retreats of the sea extended or restricted the continental land area. Episodes during which sediments were folded and carried to high altitudes—this is called orogenesis, or mountain building—were superimposed on this overall scenario. According to the traditional view, these movements were essentially vertical, horizontal displacements being restricted to between a few dozen and a few hundred kilometers.

Because their duration is so short, volcanic eruptions and earthquakes were considered autonomous events that have little relation to other objects of study in the earth sciences. The internal activity of the planet, of which earthquakes and volcanoes are the most spectacular manifestations, was decoupled from what took place on the surface. Each level, each depth was assumed to obey its own particular laws. Surface geologic activities such as erosion, sedimentation, and shoreline changes were thought to have their own causes. The activity of the deeper layers, the area called the "mantle," was thought to be governed by obscure laws in which temperature and pressure no doubt played the decisive roles and whose surface effects were limited. These laws were unrelated to the phenomena of marine transgression and regression and only slightly related to orogenesis.

Continental drift and plate tectonics changed all that. The ancient shields were no longer fixed in relation to one another, shorelines moved with their continents, successive geographies changed not only as a result of the sea's advances and retreats but also because the very framework of these movements was constantly changing. Earthquakes, volcanoes, and mountain building were directly connected to the continuous but variable mobility of the earth's surface. Earth was not a dead planet whose only geologic activity was engendered by the presence of water on its surface, but a dynamic and evolving planet whose constantly changing surface appearance, distribution of land and sea, heights and depths, and archipelagoes and platforms were the surface reflections of large-scale movements that animated its depths.

The earth is now seen as a *system* in the modern sense of systems logic; its dynamics are regulated by multiple interconnected and interregulated causes, and its behavior is as complex and global as that of a living being. Plate tectonics not only had implications for the study of causes and effects, it also furnished concrete answers to a whole series of questions that have been asked since we began observing the planet. Why are there active volcanoes in Iceland, Japan, and Indonesia, but not in Yugoslavia or Siberia? Why are California, Yugoslavia, Japan, and Indonesia ravaged by earthquakes while central and south Africa are exempt from them? Why is oil found both in Venezuela and in the Gulf of Guinea? Why is the Andean Cordillera situated along the edge of a continent while the Himalaya is located in the center of Asia? Why are there great trenches, such as those near the Puerto Rico and Kurile islands, more than 10 kilometers deep in the ocean floor? Why are the sedimentary layers deposited in Brazil so similar to those in Angola? I could fill pages and pages of equally fundamental questions that have been given clear answers since the advent of the theory of global tectonics. Furthermore, the influence of these new theories has changed the attitudes of the earth scientists and, in that way, their very organization.

The term *earth sciences*—it is important to stress the plural— refers to the totality of disciplines whose object of study is our planet; it emphasizes the multiplicity and diversity of disciplines, a vestige of the neat divisions that existed before the development of mobility theory. Before 1970 each earth scientist devoted himself or herself either to the study of a particular region of the earth or to a specialized method of approach: paleontology or geophysics, oceanography or geochemistry.

The geophysicist studied the earth's interior using the methods, tools, and measurements of physics. Since the interior offered only brief episodic exterior signs (earthquakes and volcanoes), suggesting a sort of quiescence interrupted by fits of violent activity, the geophysicist preferred to study the structure of the depths rather than their movements, their anatomy rather than their physiology. Thus fields such as seismology or gravimetry developed, specialties whose object was to determine the internal structure of the earth. Persuaded that the surface structures in no way reflected the deep structures, geophysicists remained indifferent to geology. Laboratories for geophysical research often belonged to applied physics departments, not to those of the earth sciences. The geologist studied rock formations, their distribution, and their history but considered his work isolated from that of the geophysicist. The oceanographer believed his work to be unconnected to that of the pedestrian terrestrial geologist. So continental geology and ocean geology developed in parallel, without any real links between them. Even within geology the different specialties tended to be dissociated from one another. The petrologist who studied rocks and their mode of formation did not feel concerned with the problems of marine transgressions and regressions and therefore neglected stratigraphic geology. The paleontologist who studied the evolution of life on earth considered his progress quite independent of other aspects of geology. Only the tectonic geologist, preoccupied by the difficult problem of mountain ranges, tended to use the results of other disciplines. Advances in research created new specialties, which in turn became isolated, and the tree of geologic science increasingly tended to branch out.

Because mobility theory furnished a framework for many of the great questions in the earth sciences and because it was constructed from varied and distant disciplines, it was immediately able to reconcile all the fields within the earth sciences. A phase of fragmentation was followed by one of unification and accretion. Earth scientists who thought they were investigating separate problems suddenly realized that they were all studying a common subject, the earth. In the same way that a biologist's subject is not limnology, elephants, or bacteria, but *life,* that of the geologist is not fossils, the Alps, paleomagnetism, or granite, but the *earth.* Thus a mosaic of specialists who had ignored one another became a group of earth scientists each of whom had his own outlook but pursued the same goal: to understand the earth, its functioning, and its history.

Those who passed up thematic specialization for the sake of regional studies will be forced to realize that they have chosen the wrong scale. A region is nothing unless it takes its place in a more general framework or unless the mechanisms and structures it illustrates are significant for the functioning of the earth as a whole. So, in a draconian intellectual revolution, the regional catalogue gave way to the region as a model or symbol, an example of a phenomenon. Regionalists will be even more upset when they perceive that participation in the common progress requires each person to keep his specialty, of course, but also to know enough about the others to enter into dialogues and cooperative efforts. No more living in an ivory tower! The magnetician realizes that micropaleontologists exist and that he should work with them to date the ocean floors. The petrologist perceives that unless he has some knowledge of geothermal phenomena, unless he is able to understand the geophysicist, he cannot decipher the messages in the rocks he is studying. The tectonophysicist who specializes in mountains understands that he cannot ignore discoveries about the ocean or the results obtained by oceanographers, and so forth. Thus a new, multidisciplinary scientific community is built up, one that brings together specialists who use different techniques but whose approaches and goals are more and more homogeneous.

In the midst of this intellectual banquet geology was to accomplish more in ten years than it had in the previous one hundred years; it experienced an incredible acceleration in development, a sudden crystallization of ideas that had ripened slowly, often without being formulated, and a restructuring into new chapters.

Some geologists, such as J. Tuzo Wilson, have spoken of this development, which in a few years has brought a bountiful harvest of scientific results and metamorphosed the community of the earth sciences, as a revolution. It is certainly that. The earth sciences were as radically transformed by global tectonics as chemistry by the atomic theory, physics by quantum mechanics, or biology by the theory of evolution and, more recently, by molecular biology. The appearance of mobility theory is no doubt a key moment in the history of scientific thought. This great leap forward took place in a few decisive years, but progress continues with undiminished vigor to this day.

In this book I attempt to trace in an uncomplicated manner the evolution of the ideas connected with continental mobility. This object is combined inextricably with the goal of explaining mobilist geology and its essential concepts. I have purposely simplified the work to give it a manageable size, make it accessible to the general reader, and place it in its proper sociological context. No one can understand the great adventure of our time, the development of the sciences, without also comprehending the context in which it occurs.

THE BEHAVIOR OF THE EARTH

THE WEGENERIAN SYNTHESIS

In 1912 the German meteorologist Alfred Wegener, struck by the similarity of the shapes of the coastlines of Africa and South America, proposed the theory of continental drift. In a paper entitled "Die Enstehung der Kontinente und Ozeane" (The origin of continents and oceans), Wegener suggested that Africa and South America, formerly a single block, broke apart, the two halves moving away from each other and leaving a space to be filled by the ocean. Thus, he said, the Atlantic Ocean was created by continental drift.

Wegener was not the first to state this idea. It was set forth in 1858 by Antonio Snider-Pellegrini in *La Création et ses mystères dévoilés* (The creation and its mysteries explained). After that several authors, especially Frank B. Taylor of the United States, took up the idea, but without attracting much notice in the scientific world. Wegener's contribution was to substantiate the hypothesis with the scientific arguments needed to overcome the skepticism that greets any new idea. He transformed a working hypothesis into a compact, coherent, and synthetic body of doctrine that encompassed extremely diverse aspects of earth's history in a global vision. He defended his theory firmly, but without pugnacity, until the end of his life. Wegener, therefore, must be considered the true father of the theory of continental drift. As the French historian Georges Duby said, in questions of precedence one ought to distinguish clearly between an idea put forth more or less casually among others and a work that has been constructed, argued, and developed around an idea. The first is anecdotal, whereas the second is central and obligatory.* I will follow this rule.

Isostasy

In Wegener's time the great Austrian geologist Eduard Suess, in an immense work entitled *Das Antlitz der Erde* (The face of the

---

* Georges Duby, *The Three Orders: Feudal Society Imagined,* trans. Arthur Goldhammer (Chicago, 1980).

earth), had popularized the idea that the continents, made of light granite (density 2.8 grams per cubic centimeter), "float" on underlying denser (density 3.2 grams per cubic centimeter) and more viscous basaltic rock that forms the ocean bed. Because the granitic rock is rich in silicon and aluminum, Suess called that layer SIAL; the silicon- and magnesium-rich basaltic rock he called SIMA. Like icebergs floating in water, the sialic continents are in equilibrium on the SIMA. According to Archimedes' principle, the pieces of SIAL move vertically in hydrostatic equilibrium with the SIMA. When erosion removes a surface layer of a continent, the continent tends to rise, as a barge rides higher in the water when it is unloaded. This theory, which had been developed by John Henry Pratt, Clarence Dutton, and George Airy in the nineteenth century, is known as *isostasy*.

The isostasy of the crust can be proved in various ways. Let us imagine that we remove all the water from the oceans. If we plot

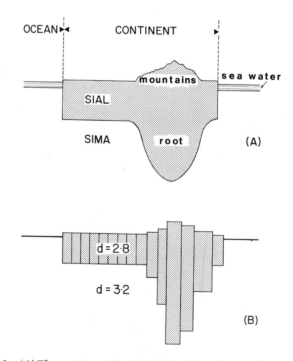

FIGURE 1 (A) The continent (SIAL) is lighter than the mantle (SIMA) and thus floats upon it. The weight of mountains is supported by a "root" of thickened continent. (B) Model of the heights of continental blocks (of density 2.8 grams per cubic centimeter) floating on a liquid mantle (density 3.2).

THE WEGENERIAN SYNTHESIS

the percent of the earth's surface that is at a given height above or below sea level, we get a curve with two maxima corresponding to the hydrostatic equilibrium level for each of the two materials, the SIAL and the SIMA: one at 100 meters above, the other at 4,500 meters below sea level. The continents play the role of the lighter iceberg.

Proof also comes from observations made by numerous geologists, including Gerard Jacob de Geer of Sweden, on the rise of the Scandinavian shield. In the Pleistocene epoch (12,000 years ago) Scandinavia was covered by an ice cap. Since then the glacial covering has retreated northward for climatic reasons, thereby lightening the shield. We know that the shield has risen little by little, because the altitude of known points on the shield has increased over time. This rising motion recalls the image of the floating barge: lightened by the melting of the ice, the shield is returning to its original level.

Finally, a more complex study of the variation of surface gravity shows that mountain ranges, like icebergs, have deep roots that counterbalance their high elevations. This too confirms the analogy between a continent floating on the ocean bed and an iceberg floating on the ocean.

Wegener supported his theory by invoking the principle of

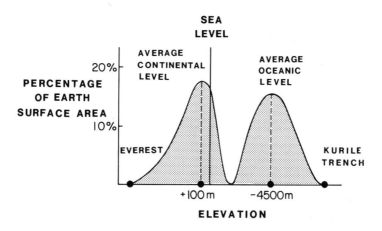

FIGURE 2   Hypsometric curve plotting the percentage of the earth's surface (with the oceans removed) that lies a given distance above or below sea level. The graph has two peaks—one at 100 meters above sea level (reflecting the average height of continents), the other at 4500 meters below sea level (showing the average depth of the oceans)—because continental and oceanic materials have different densities.

isostasy and asked: if vertical movements are possible for the continents, why not horizontal displacements? Icebergs in water are not stationary with respect to one another. Why couldn't Africa and South America have moved progressively away from each other? And why restrict continental drift to the South Atlantic? Wegener proposed to apply the theory to all continents.

Wegener's Scenario for Continental Drift

At the end of the Carboniferous period, about 270 million years before the present (M.Y.B.P.), there existed a single continent named Pangaea. In the following period, the Permian, this supercontinent

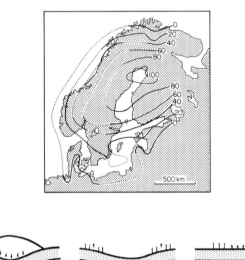

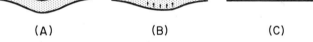

(A)   (B)   (C)

FIGURE 3   The Scandinavian shield has been rising since the removal of a giant ice sheet that covered the area during the last ice age (contour lines show centimeters of uplift per hundred years). The center of the shield is rising at a faster rate than the edges because the center region was pushed down farthest by the ice sheet. Suppose a circular ice sheet forms on the elastic Scandinavian crust. The weight of the ice will push down the crust (A). When the ice is removed (B), the crust will elastically rebound to its original height (C). Because the underlying mantle is a viscous fluid, the rebound occurs gradually and thus is still observable today.

broke apart, its pieces moving away from one another. After millions of years, in the Eocene epoch (50 M.Y.B.P.), a Eurasian continent, attached to North America through Scandinavia and Greenland, formed the northern supercontinent, Laurasia. In the south a series of blocks, called Gondwanaland, contained South America, Antarctica, Australia, and Africa, which, although it was still attached to Asia, had begun to separate from it by means of the Mediterranean Sea. More recently Eurasia moved completely away from Africa. The Atlantic, Indian, and Arctic oceans are the result of continental migrations.

For Wegener these continental movements did not simply represent the destruction of a supercontinent, they were the driving force of great geologic phenomena. The drift of continental rafts manifests itself geologically in what he called "bow and stern effects." Gigantic wrinkles were pushed up on the leading edge of a drifting continent to form mountain chains. Thus the collision between the westward-moving American continent and the Pacific SIMA produced the Andes and the Rocky Mountain chains, and Australia's eastern coastal ranges were formed by the continent's eastward movement. The leading-edge folds were accompanied by important internal repercussions, which produced the intense volcanic and magmatic activities of these regions.

At the trailing edge the phenomena were no less spectacular. The drifting continents left fragments of their borders in their wake, giving birth to island chains. In its westward drift, America left behind the arc of the Antilles. Asia's drift to the northwest created the Sunda, Kurile, and Japan island arcs. In these regions too, the movement of the leading edge had repercussions in the SIMA, setting off volcanic activity and the rise of magma.

Briefly stated, such is Wegener's theory, which I will examine here using the arguments that he and his disciples accumulated over the course of his life and that he published in the final revision (1929) of his main work, *Die Enstehung der Kontinente und Ozeane.* I will not put them in the same order or give them the same degree of importance as he did, for time has emphasized some and toned down others.

Wegener's argument was not syllogistic; it did not rely on a powerful and systematic deductive logic. As is always the case in natural science, he constructed a model, a paradigm that suggested a single explanation for numerous observations made over the course of time. These observations were diverse in nature and of varying degrees of importance.

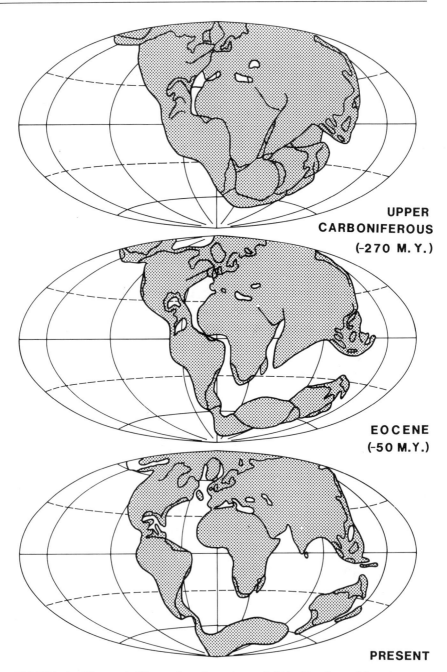

UPPER CARBONIFEROUS (-270 M.Y.)

EOCENE (-50 M.Y.)

PRESENT

FIGURE 4 Wegener's illustration of continental drift. Continental locations are shown for the present, 50 million years before the present (Eocene epoch), and 270 million years before the present (Carboniferous period). In these illustrations the African continent has been arbitrarily fixed.

Paleontology and Intercontinental Bridges

In 1912 paleontologists had already been wrestling with the problem of intercontinental contacts for a long time. Living species appeared on earth at definite times, and their remains are found in places that today are separated by oceans. Either it must be conceded that new species appear simultaneously in different places (Karl Naegeli's theory of hologenesis), or else the species must have appeared at one location and then spread throughout the geographical area accessible to it. Paleontologists rallied to the second hypothesis. But then how to explain the fact that fossils of the stegosaurus, an amphibious reptile of the Carboniferous (280 M.Y.B.P.) and the Permian (210 M.Y.B.P.) that lived along river banks and did not move about much, were found simultaneously in Europe, America (Texas), India, and South America? Examples of this sort may be found for many terrestrial species, such as garden snails, earthworms, and insects that appeared at a given time and a single place and rapidly invaded all the continents. The widespread distribution of such species led to the idea that previous connections among continents permitted the migration of flora and fauna.

But, far from proposing the idea of continental drift, paleontologists first suggested *intercontinental bridges*, tongues of land

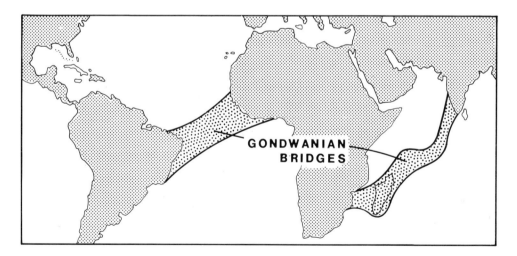

FIGURE 5   Land bridges used to be the paleontologists' explanation for the geographic dispersion of species across regions that are now deep oceans. Shown here are the land bridges proposed in Wegener's time.

linking the continents that *collapsed* in one place at a certain time, *rose* again somewhere else, and formed a vast network of intercontinental communication on a worldwide scale. Wegener attacked the theory of intercontinental bridges, because he knew that it was physically impossible for the SIAL to be swallowed up by the SIMA. The SIAL is lighter; it floats. No physical principle can account for its spontaneous sinking. Moreover, if sinking had taken place, it would have left traces in the field of gravity. Gravimetricians, however, had detected no such thing in their measurements in the oceans. Wegener saw continental drift as a synthesis of the idea of intercontinental links proposed by the biologists and that of the constancy of the ratio of land area to ocean proposed by geologists and other supporters of the theory of isostasy.

Once the theory of continental drift was accepted, paleontology became a powerful tool for reconstructing the kinematics of landmass movement. It is well known that animal and plant species evolve and change over time. On the basis of the distribution of fossils paleontologists may assign a "geologic life span" to a species. The successions of flora and fauna allow us to define the geologic time scale of epochs and eras, and the map of the range of a given species, superimposed on the map of continental drift, allows us to determine the geologic age of a given stage of continental movement. Paleobiogeographic maps could also be used to study regions for which Wegener's reconstruction was not very precise. For example, the plant *Glossopteris* was found in Gondwanaland in the Triassic but was unknown in Laurasia. Its discovery on a fragment of that continent showed that Laurasia was attached to Gondwanaland in the Triassic. Similarly, the fossil reptile *Mesosaurus* was limited to two restricted areas in South America and South Africa. From this coincidence a link between the two regions in the Mesozoic can be inferred. Using the methods of stratigraphic paleontology, which had been worked out by nineteenth-century geologists in a long and painstaking effort, Wegener established the chronology of the breakup of Pangaea and the drifting of the continents that followed it, verifying the consistency of his model in this way.

The Two Coasts of the Atlantic

Regional geology, which is based on the mapping of geologic formations, delineates the boundaries between formations. From such boundaries—for example, that between a sedimentary rock cover and a basement consisting of folded and hardened rock, which is called the basement-cover boundary—we may find evidence of continental drift. A map of the geologic formations of West Africa at its Atlantic coast poses a problem to the geologist, for some of the boundaries are practically perpendicular to the ocean. How do they continue under water? Why do they stop so abruptly?

When the jigsaw puzzle of the South Atlantic is put together using pieces shaped like its coasts, it is clear that the boundaries between the bedrock and the covering rock match. But the match is even better than it at first appears. Sedimentary rock layers are found in sequence, and they can be dated by means of the fossils they contain. Descriptions of the strata on both sides of the South Atlantic show that the layers that are older than the Triassic (that is, more than 200 million years old) are identical, stratum for stratum. The similarity is so perfect that when the South African geologist Alexander Du Toit traveled to South America in 1927 to compare the Brazilian geologic formations with those of Africa, he wrote that "formations along the two opposed shores tend to resemble one another more closely than either one or both of their actual and visible extensions within the respective continents."[*]

After making an extremely careful comparison of the correspondences among the various geological formations on both sides of the Atlantic, Du Toit extended his meticulous reconstruction to all of Gondwanaland. Further evidence came from Émile Argand, a tectonophysicist, who studied the distribution of folded mountain ranges. He too concluded that Wegener's model offered the most consistent and harmonious picture of this distribution and that nothing in the cartographic or tectonic data contradicted it.

Ancient Climates

A meteorologist by profession, Wegener was particularly interested in ancient climates. Geologic formations contain clues from which it is possible to reconstruct their history. For example, fossil corals imply warm, well-aerated waters; abundant fossil plants with giant leaves imply tropical conditions. An accumulation of striated pebbles, called glacial till, attests to the former presence of glaciers. Studying the distribution of glaciers, Wegener noticed that in the Carboniferous period South America, southern Africa, India, and Australia were covered with an ice cap. How can such a great extension of the ice cap be explained? Was the climate of the whole

[*] Alexander Du Toit, *A Geological Comparison of South America with South Africa*, Carnegie Institution of Washington Publication No. 381 (1927), p. 109.

earth colder at that time? That would be inconsistent with the presence in the same period of tropical animals and plants in the region of the present-day Mediterranean. The solution lies in the proposition that the continents were a single landmass and that the south pole was in the Indian Ocean at that time.

Another observation supports the same conclusion. The island of Timor, which lies off the coast of Australia, was being surrounded by coral reefs while the Carboniferous ice cap covered Australia. The only way to explain the simultaneous existence of such disparate climates is to postulate that the two pieces of land must have been far apart in the Carboniferous and that they have moved closer together since then. Pushing his analysis much further and suggesting the displacement of climatic zones, Wegener proposed that the poles migrated over the course of geologic time. This idea would reappear later on in the geologic literature.

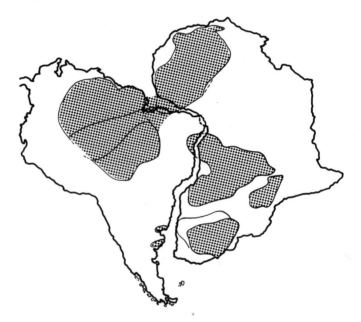

FIGURE 6 Geologic formations can be divided into two broad categories: young (less than 500 million years old), sedimentary, fossil-bearing terrains and ancient (Precambrian), crystalline, basement terrains that contain no fossils and are often folded, metamorphosed, and intruded by granites. The boundaries between these two types of terrain are easy to distinguish and thus easy to map. The regions of crystalline basement in Africa and South America (shaded area) match up on both sides of what is now the Atlantic Ocean. (Figure originally drawn by Wegener and redrawn with modern data by Patrick M. Hurley.)

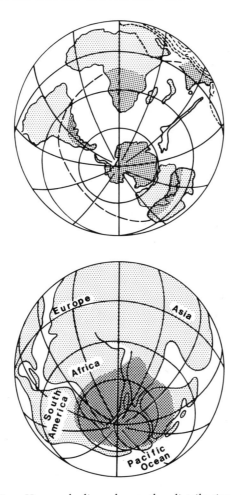

FIGURE 7 *Top:* Heavy shading shows the distribution on present-day continents of glacial deposits during the Permian and Carboniferous periods. *Bottom:* On the reconstructed continent of Gondwana, these deposits form a disklike ice sheet comparable in size to ice sheets formed during recent ice ages. (After A. Wegener.)

Mountains

In the nineteenth century geologists focused on the study of stratigraphy and the famous debates of William Buckland, Charles Lyell, Roderick Impey Murchison, Georges Cuvier, Adam Sedgwick, and Alexandre Brongniart. The goal of stratigraphy, the study of successive rock strata, is to decipher the history of sedimentary basins: geologic beds are laid down in sequence, the oldest on the bottom, and time is recorded in a series of simple messages indicating the conditions under which the rock was deposited under water. The twentieth century, on the other hand, is the era of tectonics, in which we attempt to decipher the message contained in folded chains of mountains. How were these folds formed? How did mountains as imposing as the Alps or the Himalaya arise from sediments deposited under the sea?

The study of mountains necessitates a very different turn of mind from that needed for the traditional stratigraphic approach. One requirement is the willpower to attack a problem whose complexity seems to defy solution. Also, one must be able to envision three-dimensional formations to study bold mountainous reliefs. Moreover, the study of these complex structures, with their folds, faults, and infinitely varying combinations thereof, leads immediately to a dynamic interpretation. The formation of ranges makes sense to tectonic geologists only when they allow themselves to visualize the movements that gave birth to mountains. Let us put Wegener's theory into the context of these ideas.

At the turn of the century the most popular explanation for the formation of mountains was the "baked apple theory" of the French scientist Élie de Beaumont. As an apple is baked, it loses water and contracts; its surface becomes cracked, wrinkled, and folded. An observant eye can distinguish the equivalent of mountain ranges, valleys, and ocean beds on it. In the same way the hot earth was thought to have contracted over the course of time. Adopted by Suess, this theory would have adherents until the 1960s. In the early 1870s, however, the alpine geologists Marcel Bertrand, Hans Schardt, Maurice Lugeon, and especially Rudolf Staub and Franz Kossmat called it into question by pointing out the greatly reduced land area of mountains. For example, if the Alps were "unfolded," their surface area would increase four- or fivefold. These "shortenings" imply horizontal motion, for there is no physical mechanism by which internal contraction can produce such effects. Kossmat wrote: "The formation of mountains must be explained by large tangential movements of the crust, which cannot be incorporated into the contraction theory."*

Émile Argand of Switzerland linked the tectonic geologists' "tangential thought" with Wegener's idea of continental drift. First

* Franz Kossmat, "Erörterungen zu A. Wegeners Theorie der Kontinentalverschiebungen," *Zeitschrift der Gesellschaft für Erdkunde zu Berlin*, 1921, no. 3–4, pp. 103–110.

Argand completed a great cartographic task, the tectonic mapping of Eurasia, and presented it to the Geologic Congress in Liège in 1924. In the astonishingly imagistic language that was typical of the time, he proposed that the mountain ranges from the Himalaya to the Swiss and French Alps were the result of extensive lateral movements that caused parts of the crust (the SIAL) to overlap and form the high peaks that characterize these ranges. He suggested that there was some plasticity, localized in weak areas, in the interior of Wegener's sialic blocks that allowed the blocks to be folded over and over upon one another. Africa, drifting toward Europe, gave birth to the Alps in this way. The impact of the Indian shield on the Asian shield gave birth to the Himalaya. Argand showed that in Laurasia there were rigid blocks that could not change shape and zones in which deformation was possible (although his belief that India was part of Laurasia was later proved wrong). In effect, Argand developed the idea that collisions created intracontinental mountain ranges. He compared these ranges with the circum-Pacific mountain ranges, such as the Andes or the Japanese ranges, and asserted that the ranges situated on the edge of the Pacific show much less shortening and therefore must have a different origin. This distinction became one of the foundations of modern tectonics and would reappear later in connection with plate geology.

The Development of Global Tectonics

Wegener saw immediately how to use Argand's work. He adopted it and extended it to develop the principles of what is now called global tectonics. First he established the link between global movement and regional tectonics and noted that if there is compression in one part of the globe, there must be a zone of extension or stretching somewhere else. From that basis he categorized three types of tectonic zones:

1. In *compression* zones, characterized by folds and thrust faults, the terrestrial surface is decreasing. An example is the collision-caused mountain ranges described above.

2. In zones characterized by *stretching*, "normal" faults are dominant and surface is being created. Wegener studied at length the region of great African lakes that traverse East Africa from south to north, from Lake Tanganyika to the Afar triangle, near Djibouti. He suggested that this area represented either the stage

immediately preceding the breakup of a continent (in which case East Africa will be detached from the African continent several million years from now) or, inversely, an incipient breakup that failed to develop. The phenomena that are taking place there must be similar to those that occurred in the central valley when South America and Africa began to separate from each other.

3.   The third type of tectonic zone is that in which continental movement occurs in a direction *parallel* to the faults and in which, consequently, relative lateral displacement is important. In discerning this type Wegener recognized the great impact of strike-slip faults, such as the San Andreas Fault in California, site of large earthquakes (notably the Great Quake of April 18, 1906, in San Francisco). With much foresight he drew the connection between these great shifts and the distribution and frequency of earthquakes under the various conditions of global dynamics.

Oceanic islands also attracted Wegener's attention, at first as evidence of drift. He had correctly perceived that the Seychelle Islands, in the middle of the Indian Ocean, were granitic in nature and thus of continental origin, and that the presence of a piece of continent in the middle of the ocean needed explanation. But he had generalized from this too quickly and concluded that the Atlantic islands, such as the Azores and the Canaries, were also splinters of continents. Today that is no longer believed. The shapes of the island arcs—those between Tierra del Fuego and Antarctica (the Sandwich Islands), say, or those around New Guinea—suggested to Wegener the kind of scrolling patterns observed when a liquid encounters an obstacle. The attention he gave to the birth of the island arcs left behind by Asia in its drift, for which he proposed a mechanism of creation, was more pro-phetic. At the same time he explained the birth of what would later be called marginal seas—the Japan, Okhotsk, and Bering, for example. The spatial distribution of active volcanism seemed to Wegener to testify to the existence of continental drift. The "bows and sterns" of the drifting rafts (the Andes or Indonesia) and the regions in which continents were splitting apart (great African lakes) appeared to be marked by an intense volcanism, indicative of considerable internal activity.

Thus it can be said that Wegener understood that the earth possessed a *unified global organization* and that this organization must govern a variety of diverse scientific disciplines: volcanology, paleontology, tectonics, paleoclimatology, geophysics. In this connection it is exciting to realize that Wegener devoted the last chapter of his last revision of his major work to oceanography, which was then in its infancy. He discussed topography and the distribution of oceanic ridges and commented on the contents of the first ocean dredges that brought basalt to the surface, as well as on the first geophysical measurements taken at sea. Did he envision that a striking confirmation of his views would come from these studies thirty years later?

Rereading Wegener's last pages today gives a sense of the prophetic lucidity of this meteorologist from Marburg who was such a great geologist. Why resist the temptation to give him the title his contemporaries refused him?

Causes of Continental Drift

Having proposed his model, Wegener was compelled to explain how it worked. What forces caused these majestic movements? To the naturalist his set of arguments was very convincing; to the physicist, it was less so. Even the concept of isostasy, the centerpiece of Wegener's explanation, was suspect in the eyes of orthodox physicists. The mantle consists of rigid rock under high pressure. How could Wegener compare it to a fluid, how could he speak of icebergs (SIAL) floating on water (SIMA)? For many physicists the idea was a physical absurdity, because seismology showed evidence that the earth was elastic. The interior of the earth transmits seismic waves made by earthquakes, so the earth's interior must be solid. A solid is not a fluid! Referring to Suess, Wegener invoked the factors of time and temperature, and he recalled that sealing wax is brittle if broken suddenly but is subject to plastic deformation if force is applied to it over a longer period. He had no more luck than Suess had in convincing the physicists of the time, who were more rigid than rigorous.

Then Wegener attacked the problem of the forces that might be capable of producing horizontal displacements of continents. Of necessity these forces would have to be enormous: what could they be? As a meteorologist he was interested in the earth's rotation, the Coriolis force connected with it, tidal forces, and accelerations due to the rotation. Unfortunately, when he tried to buttress these theoretical possibilities with quantitative data, he found none of these forces large enough to explain the migration of continents. Toward the end of his life, however, he envisioned the solution

Alfred Wegener

that would be accepted thirty years later: convection in the mantle.

Radioactivity, discovered by Antoine-Henri Becquerel in 1896, was still a new idea in the earth sciences in 1920 (indeed, it would still be new in 1950!). John Joly of Ireland, noting that radioactivity dissipates both heat and radioactive elements (such as uranium or thorium) in terrestrial rocks, asserted that a great deal of heat must be produced in the interior of the earth. The heat is not distributed in a uniform way and therefore induces convective currents, movements of masses of material in the interior of the earth analogous to the currents one sees in a pot of boiling water. Convection currents could move the continents floating on the viscous mantle and thereby cause continental drift.

Today no one doubts the idea of convection currents, but earlier in the century geophysicists fought it to their last breath. How could a quasi-solid substance act like a fluid? Physical nonsense! And how could the effects of radioactivity be quantified when the nature and composition of the materials in the mantle were entirely unknown? A physics of the whole earth, so different from the comfortable laboratory physics to which they were accustomed, seemed totally foreign to traditional physicists.

The Debate

Now that we are acquainted with the profusion of observations and the sparseness of theoretical interpretations advanced by Wegener, we must look at the way the scientific community received his theory.

Wegener gave his ideas their first public presentation in Marburg in 1912, and his first article was published at the end of the same year. The initial reactions of the German geophysicists were very favorable, but those of the geologists were extremely cautious, even hostile. It was not until after World War I, ten years later, that the theory of continental drift spread beyond Germany, in particular in the Anglo-Saxon world. It immediately provoked leading figures in the earth sciences to take very definite stands, and over time hostility to the idea became entrenched. When Wegener died in 1930 on one of the expeditions to Greenland that he loved so much, a good number of champions of the theory still existed. A few years later there would be only a handful left. The geophysicists, who had been temporarily attracted to the theory, ended up following Harold Jeffreys in condemning it in the name of scientific rigor and mathematics!

A "summary" of the debate over Wegener's theory emerged at the symposium on continental drift held by the American Association of Petroleum Geologists in New York in 1926. Wegener himself was absent. Of the fourteen oral presentations at the colloquium, seven were unalterably opposed to Wegener and seven were in favor (perhaps the symposium was planned that way?). We know Wegener's arguments. What did his opponents say? To summarize the most important objections:

Wegener's paleogeophysical reconstructions were only approximate. Although the South Atlantic coastlines seemed to fit together properly when the existence of a continental shelf was taken into account, the same could not be said for the North Atlantic. How could Newfoundland be fitted into the coast of Europe?

How could geological features on both sides of the Atlantic fit together so well when the continents have supposedly undergone a long trip that would be expected to have distorted them? The existence of a good match is proof of the *immobility* of structures! That argument, which would reappear in a more "rigorous" mathematical form proposed by Jeffreys, would prove to be one of the deadliest for the drift theory.

Why did the splitting up of the continents not begin until the Permian period? Why did Pangaea survive throughout the major part of earth's history? The idea that mobility was a recent phenomenon seemed to contradict the popular idea that the earth was much more "active" in its early history than at present. This argument would reappear later, also.

Among the other objections to Wegener's theory were the absence of "convincing forces" to propel continental drift and the peculiar physics needed to explain how rocks in the earth's interior could behave like a fluid under the action of temperature, pressure, and time. In addition, the idea of folded mountain ranges caused problems. If the mantle constitutes the plastic part of the globe, why is it that the superficial continental crust does the folding to give birth to mountains?

The president of the symposium, Willem van Waterschoot van der Gracht of Holland, was remarkably perceptive in his closing remarks. He emphasized that the theory of continental drift accounted for many paleontological, stratigraphic, and climatological facts but that no physical theory explained the observed phenomena.

Reading the communications of the symposium today, one is surprised by the harshness of the criticism of Wegener and his ideas—they were called "pseudoscience," "superficial approach," and "manipulation of objective facts." At the same time, an examination of the objections of the opposition reveals an extraordinary lack of substance. Except for those who focused on the inadequacy of Wegener's causal explanation, none of the critics dared to attack the basis of the reasoning founded on geological, paleontological, or tectonic observations. Neither did anyone propose another solution or an alternative hypothesis. Wegener's opponents destroyed a splendid synthesis and replaced it only by chaos.

One might wonder why a theory whose beautifully synthesized form contrasted strongly with the sparse and ill-assorted facts of the earth sciences of that time—and which was, as we know today, essentially correct—did not take hold in the scientific world. The answer is a difficult one, for it arises from what could be called the history of the scientific mentality. One could argue that if Wegener had presented his theory in China at that time, it would have received a much warmer welcome. Chinese scientific thinking, infused with Taoism, was concerned with erecting a hierarchy of nature according to a synthetic classification based on objective observation; it would no doubt have paid little attention to the weakness of the physical proofs. Western thought, on the other hand, dominated by Greek philosophy and especially by the principle of cause and effect, has always had a great deal of trouble grasping a physical problem whose causes cannot be perceived. It can easily accommodate an imperfect understanding of those causes, but not their total absence. The result is the cult of the theoretician who explains things, as opposed to the observer who describes them. Perhaps the rejection of Wegener's theory originated from such a state of mind.

Other explanations can be found in the psychosociology of science. In an anecdotal and encyclopedic science like traditional geology, knowledge and competence gained little from deductive power and a great deal from memory and experience. Thus only an experienced scientist could achieve the stature of the "learned man," a term which is consciously or unconsciously confused in people's minds with a certain maturity and extensive experience. How could a neophyte, someone who was not even a geologist, find the key to problems that experienced scholars had been studying for such a long time? The theory of continental drift seemed much too easy in an era in which hard work and effort were seen as the ideal. Perhaps also the idea that every bit of earth had a single root in one primitive continent, and that distinct geographic areas evolved from a common origin, suggested the idea of uniformitarianism. This idea could be seen as a resurgence of the Darwinian theory of evolution, which was then being fought vehemently. This, of course, is only a hypothesis.

No doubt Wegener's personality and position played a role also. Wegener was a meteorologist and passionate devotee of Greenland exploration. He expounded and defended his geological theory, but without pushing it to the limit, without vehemence. There was no boisterous correspondence, no spectacular outbursts. Wegener did not belong to the geological "establishment"of the time; he did not teach geology or practice it in the usual sense of the term. He was a loner, an outsider, and he therefore appeared to be an astute but not a credible amateur.

No doubt all these factors played a role, but perhaps Wegener was really defeated by the fear of change, the force of habit, the corrosive skepticism that caused one participant in the New York symposium to say: "If Wegener were right, gentlemen, we would have no choice but to go back to our schoolbooks . . ."*

It took forty years for the geologic and geophysical community to consent to that!

* The papers of the symposium were published in W. A. J. M. Van Waterschoot van der Gracht et al., *Theory of Continental Drift: A Symposium* (Tulsa: American Association of Petroleum Geologists, 1928).

RETREAT TO SPECIALIZATION 2

AFTER Wegener's death and despite the efforts of scientists such as Du Toit of South Africa, Arthur Holmes of England, and B. Choubert of France, continental drift and global geodynamics faded from the minds of earth science researchers. Wegener had assembled the orchestra; now it broke up: the paleontologists went back to their fossils, the geophysicists to their calculations, the geologists to their favorite area studies. Given this victory of darkness over light, of narrow-mindedness over global thinking, one might think that the period that followed would have been a particularly dark one for the earth sciences. Nothing of the sort. Many rejected Wegener's ideas out of the timidity and the parochial prejudice that characterize small minds, no matter how scientific they may be, but others scorned them in the name of scientific rigor. The latter, or rather their spiritual descendants, applied themselves to the construction of a "serious" geology built on observation, quantitative measurement, and calculation.

Although progress was slow, unspectacular, and very specialized, the development of careful scientific methods eventually produced an abundance of results. By introducing the most recent advances in physics, chemistry, and technology into the earth sciences, scientists were able to develop a series of extremely powerful tools. These developments took place in no particular pattern of method or time, each independent from the rest, and so I will present them in no particular order. They formed the foundation for the startling advances in modern earth science starting in the 1960s—and for the rebirth of Wegener's theories.

Geologic Mapping

To find valuable minerals, especially petroleum, geologists map the principal regions of the earth systematically. The association between certain geologic formations and valuable resources (basins and petroleum, granite and minerals) changed what had been mere scientific curiosity into economic necessity.

In the absence of a unifying theory for the earth sciences in the decades after the refutation of Wegener's theory, both laboratory researchers and field geologists made cartographic surveying a "fervent duty." Little by little the geologic map of the world was filled in. The task benefited considerably from the military's need for and distribution of aerial photographs. Using photographs, an experienced geologist could map a large area with fewer on-the-spot observations. With new cartographic data it became possible to piece together a picture of the whole earth with its old shields, great sedimentary basins, mountain ranges, and ancient coastlines. Moreover, the picture was a chronological one, for each layer of rock could be dated on the basis of the fossils it contained. Thus one could show a series of landscapes from past geologic eras in a series of maps; such work is the basis of *paleogeography.*

Mapping absorbed about nine-tenths of the efforts of geologists throughout the world. It was a fundamental and indispensable step, but no new concepts emerged from it. However, some scientists made what amounted to a religion out of it and still practice it without knowing why they continue to do so. As each region was more and more completely investigated, the areas mapped became smaller and smaller. The "mapping mentality" that began as a road toward synthesis often turned into a mania for more and more minute analysis. A means became an end in itself; generations of geologists wore themselves out in pursuit of it.

The thankless task of analysis also allowed "geological hegemonies" to develop. Each student was shut up in his own area; only the professor theoretically had access to the total picture, leaving the student the job of a poorly paid clerk with his nose in his account book. Thus, at the age when imagination blossoms, young geologists had no experience with synthesis; in their maturity some of them realized the possibility for analysis, but for many their imagination had dried up by that time. The accumulation of details created the desire for more details; cartographic perfection led to perfectionism. Little by little the dreams and visions of youth fade away!

Of course, some scholars resisted this tendency and proposed broad theories: first Argand in his tectonic study of Asia, and then S. Warren Carey of Australia, L. Glangeaud of France, and Al Engel of the United States. Unfortunately, their theories were not taken seriously, because they lacked concrete proofs.

The mapping of folded mountain ranges should have breathed

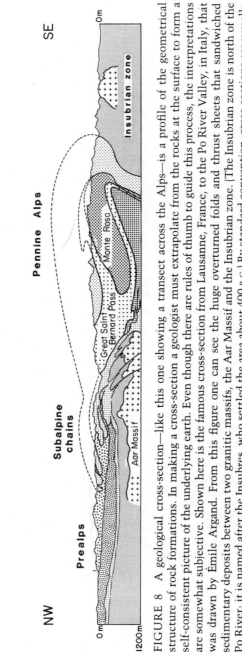

FIGURE 8   A geological cross-section—like this one showing a transect across the Alps—is a profile of the geometrical structure of rock formations. In making a cross-section a geologist must extrapolate from the rocks at the surface to form a self-consistent picture of the underlying earth. Even though there are rules of thumb to guide this process, the interpretations are somewhat subjective. Shown here is the famous cross-section from Lausanne, France, to the Po River Valley, in Italy, that was drawn by Émile Argand. From this figure one can see the huge overturned folds and thrust sheets that sandwiched sedimentary deposits between two granitic massifs, the Aar Massif and the Insubrian zone. (The Insubrian zone is north of the Po River; it is named after the Insubres, who settled the area about 400 B.C.) By standard convention, cross-sections are usually exaggerated vertically by a factor of 10–20. It is also common to show with dotted lines the rocks that have been eroded since the folding took place. (After H. and G. Termier.)

new life into the theory of mobility, for this meticulous task fully confirmed the interpretation of the pioneers of alpine studies, namely, that rock masses move laterally. The three-dimensional character of geologic maps becomes particularly clear in cross-sections. Cross-sections of the Alps show vast folded strata that are highly overturned, suggesting that squeezing has occurred. In other words, they show lateral movements that in turn suggest the existence of lateral compression forces and, therefore, of a certain mobility. Paradoxically, tectonic geologists included lateral mobility in their reasoning for thirty years without making a general principle of it or subscribing to Wegener's ideas. They would accept only a limited mobility, making no allowance for movement on a global scale. Moderation in all things!

The Seismology and Internal Structure of the Earth

Seismology is the study of acoustic waves traveling through the interior of the earth. The characteristics of wave propagation permit us to determine the physical properties of the material the waves have crossed. As X rays allow us to determine the interior structure of the human body, so, analogously, seismic waves allow us to determine the composition of the earth's interior.

In seismology an earthquake or, more rarely, an artificial explosion emits the waves. The receiver, a *seismograph*, registers vibrations in the ground and transmits the vibrations in the form of a graphic record called a *seismogram*. Seismologists can distinguish among various types of waves according to the shape of the seismogram. Observatories around the world record the times at which the various waves from a given earthquake arrive, and the path the waves followed through the earth and their speed between two given points can be determined from the arrival times.

The speed at which the waves are propagated varies according to the type of material being traversed, therefore one can make a profile of seismic speeds through the various parts of the globe. The birth of seismology coincided with the Wegenerian period. In 1883 John Milne announced that earthquakes must be "audible" all over the earth. Six years later Ernst von Rebeur-Paschwitz detected a Japanese earthquake in Germany. Between 1905 and 1915 the principles of seismology were laid down, thanks to the work of Milne, Richard D. Oldham, and Ernst Weichert. Oldham showed that seismograms consist of the superposition of trains of waves of different types. *P* waves move through the earth's interior and arrive first; *S* waves follow the same path but arrive more slowly. *Surface waves*, which follow a much longer path, arrive after the first two. The difference between the arrival times of the P and S waves at a given point makes it possible to determine the distance of the observation point from an earthquake. Milne showed that with three stations one can determine the epicenter of an earthquake. That demonstration was the kick-off for seismic cartography.

While studying the propagation of P and S waves, Oldham discovered the existence of a core at the center of the earth and in 1914 Beno Gutenberg specified its position at a depth of 2,900 kilometers. In 1909 Andreiji Mohorovičić of Yugoslavia showed that the speed of seismic waves changes abruptly when the waves reach the inner edge of the earth's crust. The place at which the speed changes is called the Mohorovičić discontinuity (or "the Moho"). Thus, in Wegener's time the major outline of the earth's internal structure—the superficial crust, the central core, and the mantle in between—was well known.

Theoretical seismology, a new branch of the physical sciences, began to develop around this structure. Its object was to determine how seismic waves are propagated through the earth and how they interact to form the complex seismograms recorded at every station. One of the basic assumptions of this field is that the earth is spherically symmetrical. The earth's surface geography is complex, but its interior consists of spherical layers that fit one into another. The properties of the mantle are almost identical 600 kilometers under Paris, Adelaide, Hawaii, or the Kerguelen Islands. A description of the average earth therefore suffices to describe the real earth.

The period of "spherical seismology," which lasted from 1915 to the Second World War, was dominated by two scientists who were as exceptional as they were dissimilar: Harold Jeffreys, the man who had "proved" the impossibility of continental drift, a brilliant English mathematician, and specialist in the calculation of the propagation of seismic waves; and Beno Gutenberg, the German founder and director of the seismology laboratory at the California Institute of Technology (Caltech), an "observatory physicist" who used a minimum of mathematical tools (which does not mean that his work was not theoretical in many ways), and a true artist at deciphering and interpreting seismograms. These two scientists, who were as different from as they were complementary to each other, developed an "average earth" model that has in general remained unaltered up to the present day.

According to this model the earth can be described as a sphere

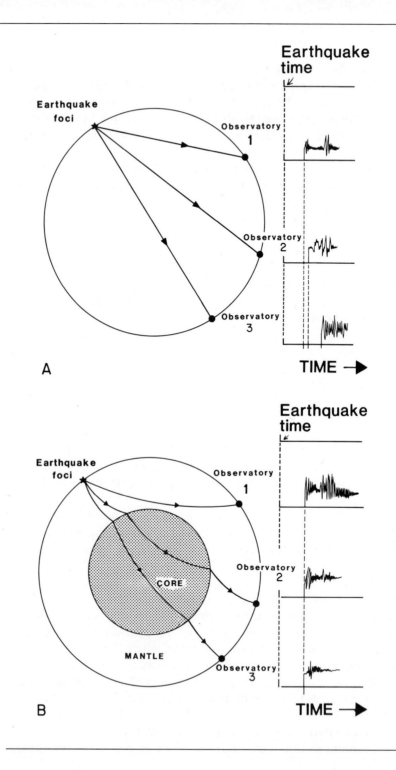

whose density increases constantly as one moves toward the center. The increase is not continuous but increases abruptly in zones a few kilometers in depth. This is the case for the boundary between the crust and the mantle (the Moho), the mantle-core boundary, and also for two discontinuities located 400 and 650 kilometers from the surface within the mantle. In 1936 Inge Lehmann of Denmark showed that the core consisted of two parts: the inner core and the outer core.

Gutenberg and Jeffreys clarified the difference between the two parts by showing that the outer core is liquid and the inner core solid. But for our purpose the most important discovery was made by Gutenberg in 1926 while he was studying an earthquake in Chile. He showed that there is a "soft" layer in the mantle, called the asthenosphere, at a depth of about a hundred kilometers. The asthenosphere would be studied in great detail by Gutenberg's successors at Caltech, especially by Don Anderson around 1960, who used long-period waves, the surface waves, for that purpose. The studies showed that the asthenosphere underlies the entire surface of the earth, both continents and oceans, and that it is both less dense and more plastic than the layers above it. The upper layers, which are rather rigid, are separated from the surface by a thin layer called the crust. So the picture that we have today of the earth's structure was born: the earth is like an egg with a shell (the crust), a white (the mantle), and a yolk (the core)—except that under the shell there is a somewhat soft zone, the asthenosphere, and that the yolk is, strangely, liquid on the outside and solid on the inside. The thicknesses of the various layers were well known at that time. They range from several dozen kilometers for the crust to 2,900 kilometers for the mantle and 3,400 kilometers for the core.

FIGURE 9 (A) Seismological model drawn under the assumption that seismic waves propagate with uniform speed through the earth. The lines show the paths of the earliest-arriving seismic waves from an earthquake occurring at the star to observatories 1, 2, and 3. If the waves travel at the same speed, the closest observatory will record the seismic signal first.

(B) In this seismological model it is assumed that waves travel at varying speeds. Seismic waves increase in speed with depth, thus the seismic ray paths are curved. In addition, the iron core of the earth transmits sound waves much faster than the silicate mantle. Rays that travel through the core of the earth, even though they travel farther, may arrive at the same time as waves that travel a shorter distance but entirely within the mantle. Because of this difference in speed, the arrival times are quite similar at observatories 1, 2, and 3.

After World War I seismology was dominated by an idea that now seems obvious but that emerged from the data only gradually: the interior of the earth has a geography of its own. With this realization began the era of *deep structural geophysics,* the basis of all understanding of the dynamics of the earth, and it has not yet ended. K. Wadati of Japan opened this era in the 1930s when he noticed that deep earthquakes are found only in certain parts of the globe. His studies demonstrated that the foci of earthquakes are localized along inclined planes in well-defined regions such as Japan, Chile, and Indonesia. Fifteen years later Hugo Benioff rediscovered these findings, and the region of deep earthquakes is rather unjustly called the Benioff zone. I will call it the Benioff-Wadati (BW) zone.

Maurice Ewing and his team at Lamont Geological Observatory showed that the thickness of the crust is not the same under the oceans as under the continents. Around 1958 Frank Press and his students at Caltech made a second important discovery: the continental crust and in particular the Mohorovičić discontinuity vary in depth from place to place. That is, geophysical formations of variable shape correspond to variable geologic formations at the earth's surface.

K. Wadati

It would be tedious to report all the work done in this vein. It is sufficient to note that the idea that the earth is *not* spherically symmetrical gradually gained ground. The change occurred slowly, for abandoning a principle like spherical symmetry involves complications both in interpreting seismograms and in making calculations, which under the new model required much greater effort. Because in those days the lengthy calculations had to be done without recourse to computers, efforts were necessarily very limited and simplified. The evolution of this idea prepared geologists and geophysicists for a dialogue by bringing seismology to light and by suggesting that geological surface features were not independent from the nature of the interior.

The development of a geography of internal structures began with the study of a series of particular zones, starting with those in which liquid is found. Actually, except for the external core, the whole earth is solid, especially the mantle. This had been demonstrated by seismologists who noticed that transverse waves, the *S* waves or shear waves, which cannot cross a liquid, propagate through the mantle. These observations destroyed the idea that has often and unfortunately been spread around, that red-hot volcanic

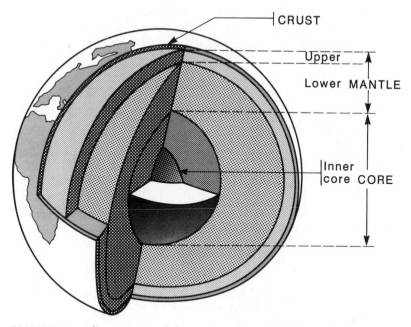

FIGURE 10   The structure of the earth as determined by seismology.

magma lies directly under our feet. But in places where there are pockets of magma, little anomalies in S-wave propagation are found also. These places are located under the oceanic ridges, island arcs, young volcanic islands, and great rift valleys (the Rhine graben, East African rift valleys) that split the continents and in which volcanic phenomena are located.

For a long time seismologists considered themselves earth physicists. They were more interested in the way waves propagate than in determining what structures they cross. The discovery of what they called "lateral heterogeneities of propagation," which complicated theoretical propagation models considerably, led them to become more interested in the structure of the earth itself. The connection between the interior and the surface, which had at first been neglected and even denied, became evident. Seismologists began playing an ever more active part in the earth sciences until eventually they took on a central role in the development of plate tectonics. But let's not skip ahead.

The Birth of Geological Oceanography

It took the Second World War to demonstrate the strategic and economic importance of oceans and submarines. How could two-thirds of the earth's surface not conceal riches as abundant as those found on the continents? These motivations were sufficient to prompt the United States, and to a lesser degree the Soviet Union, to launch a series of oceanographic ships to explore the ocean. Scientists aboard these ships, which were often quite small, made topographic profiles of the ocean floor using the sonar technique developed during the war to detect submarines. They took samples of sediments and hard rock and measured the magnetic field for each sample point, which enabled them to chart magnetic maps whose utility will be discussed below. Later, by recording small explosions on seismographs they tried to determine the character-istics of the oceanic crust and, following the example of Teddy Bullard and Roger Revelle, they measured the flux of heat escaping through the ocean floor.

Techniques for making the various measurements had to be invented, tested, and perfected during continuous functioning at sea, often under extremely harsh conditions. This necessitated a sustained effort, particularly a financial one. By 1982 it cost $22 million to fund one oceanographic research vessel, the *Glomar*

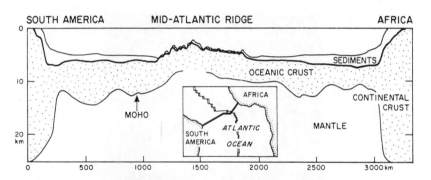

FIGURE 11   Because the continental crust is thicker than even the highest relief, in accordance with Archimedes' principle, it was therefore expected that the rising crust of the mid-oceanic ridges would also be thicker, but this was not the case. This apparent violation of Archimedes' principle was not the result of an anomalous load, and the paradox was unsolved for sixty years. It is now explained by the existence of a hot, and therefore lighter, mass in the mantle under the ridge, as Manik Talwani, Xavier Le Pichon and Maurice Ewing had proposed in 1965.

Challenger (which began operation in 1968).* Clearly, any work undertaken at sea would have to be of use to a number of scientific groups. On the other hand, a ship operates twenty-four hours a day, making practically continuous measurements and accumulating a large body of data that must be processed and interpreted. In this field the computer found its first geological application. Last but not least, specialists from many fields—geologists, sedimentologists, geophysicists—often live together and exchange ideas for several weeks on board ship in the course of a cruise. A spirit of interdisciplinary teamwork was born on those ships long before it became common elsewhere. It is therefore not surprising that oceanography played such a large part in the evolution of modern earth science.

The first of the great American oceanographic institutes to support research in the earth sciences was Lamont Geological Observatory, of Columbia University in New York City, estab-

* Sixty percent of these funds were provided by U.S. agencies, the rest by participating foreign governments. (Steering Committee for Academic Research Fleet Study, Ocean Sciences Board, Commission on Physical Sciences, Mathematics, and Research, National Research Council, *Academic Research Vessels, 1985–1990* [Washington, D.C.: National Academy Press, 1982], pp. 39–40.)

Maurice Ewing

lished on the banks of the Hudson in 1945. It was created, guided, and enlivened by the invincible tenacity and ingenuity of a most unusual man, Maurice Ewing ("Doc Ewing," as he was known to his friends). Under his firm direction the laboratory remained in the forefront of oceanographic research for twenty years. The Scripps Institution of Oceanography joined the research effort in the 1950s under the impetus of another very talented man, Roger Revelle. Its decentralized character contrasted sharply with the hierarchical form of Lamont. With Lamont scientists working in the Atlantic and Scripps personnel in the Pacific, a great deal of data was collected in a few years; emulation and sometimes competition between the two helped to keep the quality and intensity of their scientific research at a high level, but still the light of a new synthetic theory had not yet dawned. Both institutions, however, added to the store of observations.

First, five types of structures were distinguished in the topography of the ocean floor:

1. The continental slope, which follows the shallow continental shelf (a natural prolongation of the continent) and which passes abruptly (at a slope of 15 percent) to the abyssal plain (4,000 meters deep).

2. The abyssal plain, which extends over more than half the surface of the ocean floor.

3. The oceanic ridges, which are submarine mountains attaining heights of 3,000 meters and lying in long lines that run across the ocean floors in an immense network. The ridges are an important site of active submarine volcanism.

4. The gigantic submarine trenches, which lie parallel to the coast along some continental margins between the continental slope and the abyssal plain and which reach depths of 11,000 meters. A series of trenches, for example, borders the west coast of South America; another, the west side of the Pacific from Tonga to the Kurile Islands.

5. The volcanic islands, which are strewn about the oceans. The islands are particularly interesting because, although sometimes there is no apparent organization to their distribution, more often they are aligned in specific directions, more or less perpendicular to the directions of the ridges.

Next, the subsurface of the ocean floor was investigated. The propagation of seismic waves showed that the subsurface of the oceans is very different from that of the continents. Below several hundred to several thousand meters of soft sediments, the speed of *P* waves changes abruptly and reaches 5 kilometers per second, then increases slowly to a depth of 5 kilometers, where it changes abruptly to 7 or 8 kilometers per second. This transition line corresponds to the Mohorovičić discontinuity, which exists under the continents also but at a depth of about 30 kilometers. That is a primary structural difference between continents and oceans: not only is the thickness of the crust different, but so is its composition. Although the continental crust is mostly granite (a light, silica-rich rock), the oceanic crust is made of dark basaltic rock near the surface. One of those naked truths that often occur in science, truths which may seem obvious to the eyes of the uninitiated but are great discoveries to the expert, came to light: *the oceans and the continents are geologically different.* Actually, nothing suggested such a distinction a priori, because the presence or absence of water could have been by chance, as is demonstrated by the present (the English Channel) or past (Michigan Basin) existence of bodies of water on the continents. As I will point out throughout this book, the contrast between oceans and continents is probably the single most important fact in the earth sciences.

I will return often to the very costly but productive and inspiring results of these oceanographic studies. At the time the studies were undertaken, however, their importance and scientific usefulness had not yet been recognized. There was no global synthesis in which all the observations could be united; each one posed a new problem. The youth of the field and the small size of the areas explored made synthesis extremely difficult, because new data were continually calling into question interpretations of the old data. True field geologists, the continental geologists, paid little attention to ocean geology, which they said had little relation to their own. This is a theme whose echoes we will continue to hear for a long time.

The Geologic Clock

In 1896, in a building situated at the back of the Jardin des Plantes in Paris, Becquerel discovered radioactivity. Several years later, Pierre and Marie Curie isolated the elements responsible for this phenomenon, and in 1908 the great British physicist Ernest Rutherford suggested that radioactivity could be used as a geological chronometer. The basic principle Rutherford discovered is relatively simple. A radioactive isotope, called the parent isotope, disintegrates and gives birth to a nonradioactive isotope, called the daughter isotope. The quantity of radioactive isotope that disappears or of daughter isotope that appears depends only on the length of the time elapsed. The process is independent of exterior factors such as temperature, pressure, or chemical environment. By measuring the number of parent isotopes that remain and the number of daughter isotopes that have been created in a sample, we can calculate the quantity of time since the sample was formed, just as we can measure elapsed time by measuring the sand left in the top half of an hourglass or that having already fallen into the bottom half. For example, when a mineral or a rock crystallizes and "locks up" a certain quantity of radioactive isotopes (such as rubidium 87 or uranium 238 or potassium 40), it sets off the "hourglass." When we compare the quantity of strontium 87, lead 206, or argon 40 that has been produced with the remaining amount of rubidium 87, uranium 238, or potassium 40, we can calculate the time at which the rock or mineral was formed. The

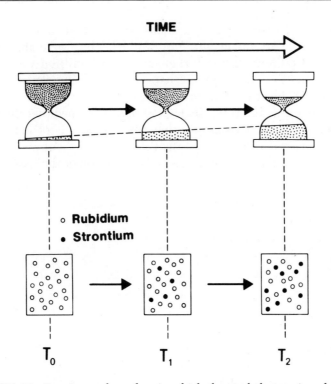

TIME

o **Rubidium**
• **Strontium**

T_0 T_1 T_2

FIGURE 12 Imagine an hourglass in which the sand changes in color as it falls. As sand falls from the top to the bottom of the hourglass, the proportion of dark sand to light sand diminishes continuously. This ratio of dark to light sand can be used to measure time. Similarly, radioactive decay of rubidium 87 atoms (open circles) into strontium 87 atoms (filled circles) depends only upon time elapsed. The ratio of rubidium to strontium atoms in a rock sample thus can be used to measure the time since the rock was formed.

isotopes used for this purpose are not ones that disintegrate rapidly in the laboratory, but those whose disintegration time is of the order of *billions* of *years.*

The principle is simple, but its application is difficult, because the interesting isotopes are found only in very small quantities: their concentration is measured in *millionths.* Therefore, highly refined techniques of physics and nuclear chemistry were needed to date rocks in this way. But, thanks to the pioneering work of scientists such as Alfred Nier and Mark Ingham of the United States, these problems were solved, and after 1937, but especially after 1948, the ages of rocks were obtained in abundance. These ages were called "absolute," as opposed to the relative ages obtained by paleontological methods. For the first time geologists were able to date the events they studied. They were able to say "This is 3 billion years old, that is only 200 million years old," rather than "This is older than that" or "This is Precambrian, that is Permian." Ideas of anteriority and posteriority were replaced by those of age and duration. Geologists rapidly learned that the earth is about 4.5 billion years old, that the first fossils of higher life forms appeared at the dawn of the Paleozoic era about 550 million years ago, and that the fossil man characteristic of the Quaternary

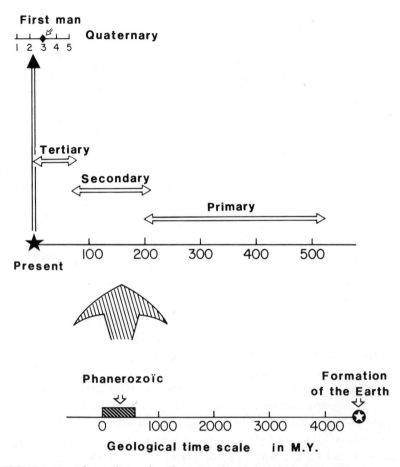

FIGURE 13   The traditional geologic epochs apply only to 500 million years of earth history and are of quite variable duration. The fossil records of earlier periods are less well preserved, so the geological eras defined on the basis of fossil evidence are longer and less differentiable.

period is only 2 million years old! Measuring time allowed geologists to measure the speed of geological phenomena, which take place over millions of years. The discovery of the geological clock marks the beginning of quantitative geology. No physics of the earth's past is possible without quantification; remember Lord Kelvin's speculations on the thermal history of the earth and its physical evolution! He estimated the earth to be between 40,000 and 4 million years old, and an intense controversy arose between him and naturalists such as Charles Darwin. Rutherford's discovery marked the definitive end of Lord Kelvin's theories.

Now it was possible to date ancient crystalline geologic formations directly and to establish simultaneities on a global scale. One could create a geology of ancient regions where traditional geology had stopped short, for lack of fossils. It also became clear that four-fifths of earth's history, the entire Precambrian era, was not covered by classical geology. Thus, exploration of the oceans allowed scientists to complete the spatial exploration of the globe, and radiochronology supplied the means of temporal exploration. Traditional geology, which had been founded on the study of fossils and sedimentary layers deposited on the continents, could account for only a third of the earth's surface and a tenth of its lifespan!

Making Rocks in the Laboratory: The Origin of Granite and Basalt

Everyone knows that the materials that constitute our planet and that one can pick up on the ground are rocks. The rocks themselves are composed of mixtures, assemblages, or, better yet, societies of minerals. Thus granite is an association of three minerals: quartz, feldspar, and mica. Basalt is composed of two principal minerals, pyroxene and plagioclase, with some olivine, the mineral that is the essential component of peridotite. The "citizens" in these societies, the minerals, are of widely variable character. Sometimes they are large in size, on the order of a centimeter, and have well-defined planar surfaces, sometimes they are small and have complex shapes and random orientation.

The study of the description, classification, and reconstruction of the formation of rocks is *petrology* (or petrography). It has long been known that there are three types of rocks. *Sedimentary rocks*, like limestone or sandstone, are formed under water in depressions or basins. *Igneous* rocks, such as basalt and granite, result from the crystallization of molten magma. They can be produced either by volcanism at the earth's surface, as basalt is, or by plutonism deep in the earth's interior, as granite is. The third type of rock, *metamorphic* rock, is more difficult to define. It results from the transformation of sedimentary or igneous rock; the new rock recrystallizes, giving birth to new associations of minerals.

The genesis of sedimentary rocks, which takes place practically before our very eyes, is hardly mysterious. The same cannot be said for igneous or metamorphic rocks, which are products of the earth's interior. How can the curious phenomenon that metamorphoses rocks take place? How are the magmas that give rise to rocks such as granite or basalt isolated, segregated, and identified? Can such magmas be created in the laboratory? Or is this the dream of an alchemist, or a curious adolescent?

In the past fifty years scientists made this dream come true with autoclaves, extremely powerful "pressure cookers." The principle of autoclaving is simple enough. A powdered mixture whose chemical composition corresponds to that of the rock one is trying to reproduce is put in a closed container and heated. The higher the temperature, the more quickly the reaction takes place and the more easily a rock is formed. But heat alone is not enough. The high pressures of the earth's interior must be reproduced, also. The simplest way of building pressure is to mix in with the powdered rock a gas that will react to the temperature increase by expanding and causing an increase in pressure. The gas may react chemically with the powder to produce minerals: this is what happens when water vapor, which is very reactive chemically, is used. On the other hand, chemically inert gases such as nitrogen and argon may be used; these do not react with the rocks with which they come in contact. In this way pressures of several million atmospheres are produced.

These experiments are limited by the capacity for resistance to great pressures of the materials of which the autoclave is made. Under such high internal pressures the instruments naturally tend to explode. The Carnegie Institution of Washington quickly became the temple of this new kind of alchemy: various minerals were synthesized under a variety of conditions of temperature (from 250 to 1,000 degrees Centigrade) and pressure (up to 50,000 atmospheres). Scientists at Carnegie were then able to synthesize rocks and to determine which associations were stable, what reactions took place among minerals, destroying some and forming others, and what conditions were necessary for chemical reaction. To complete this task fit for a Benedictine monk it was necessary to study systematically all the known mineralogical associations in the entire range of possible temperatures and pressures. It became clear that for a given chemical composition a certain mineralogical association could be formed only within a well-defined temperature and pressure zone. When a sufficient number of duly repeated and verified experiments had been performed, an elaborate table was compiled specifying a mineralogical association for each temperature and pressure zone and for each given chemical composition. This table is the *mineralogical code,* from which it is possible to tell what conditions of temperature and pressure existed in the environment at the time of the formation of a given metamorphic rock. Thus each metamorphic rock contains a message that it would henceforth be possible to decipher. In combination with radioactive dating methods, the mineralogical codes made it possible to describe the thermal history of the planet. So this was a very useful tool for geological reconstructions.

Legitimately proud of this success, experimental petrologists then tackled a more difficult problem: the birth of magmas. How do rocks melt to form magmas? As we know, granite is the essential component of the continental crust. An understanding of its birth and manner of formation, therefore, would greatly advance the understanding of the formation of the continents themselves. Naturally the alchemist-petrologists speculated on these matters. The first laboratory experiments, in the 1950s, showed that granites can be produced by chemical differentiation of basaltic magmas, a process of distillation.

Certain minerals crystallize out during each stage of the cooling of a basaltic magma. The crystals are heavier than the magma, so they fall to the bottom, leaving a liquid that is enriched in those chemical elements not yet crystallized. Because the first minerals to crystallize are "poor" in silica, they leave behind a residual liquid rich in silica. Thus, a granitic melt is formed through successive crystallizations. Discovering this process was a truly spectacular achievement, for it explained how granite is produced from a basaltic melt, the magma from which most terrestrial volcanism originates and whose source is the depths of the earth. The continental granitic crust appears to be the result of differentiation of the basaltic magma from the earth's interior.

Before the autoclave experiments it was thought that the origin of the great structural elements of the earth (core, mantle, continental crust) had been definitively explained. At the beginning of time, the earth was a molten mass. Heavy metallic iron separated and sank to the earth's center, while the light granite floated toward the surface. According to the description of this process given by the great German geochemist, Victor Goldschmidt, this differentiation is like that which takes place in a metallurgical furnace: the slag floats and the pure metal falls to the bottom.

As is often the case, one experiment was enough to call into question even the most beautiful theory. When Norman L. Bowen, one of the great pioneers in rock alchemy, and O. F. Tuttle attempted to melt sediments in the presence of water and under a pressure of several thousand atmospheres, they were surprised to find a piece of real granite in the autoclave, showing that granite could be formed by the melting and fusion of sediments. This possibility opened new perspectives to understanding how the continental crust was formed. It was possible to imagine that the sediments, a product of superficial alterations, were buried in the earth, heated and pressed together there, recrystallized to form metamorphic rock, and then, continuing their downward journey, melted to form granites.

Such a scenario was not simply a product of the geologists' imagination. It corresponded exactly to what can be seen in the valleys that cut deeply into the great mountain chains such as the Rocky Mountains, the Alps, or the Himalaya. One can see a progressive passage from sedimentary to metamorphic rocks (mica schist) and then finally to granite. The implication of this picture is that granite is the result not of vast internal differentiation but of external processes, complex events that have taken place over the course of geologic time on the earth's surface. According to the first theory, granite is the product of fire; according to the second, it is the child of water. This debate, whose origin dates to Mesopotamian cosmologies, divided the geological community for ten years without a consensus being reached. It could be said that until 1966 the majority of Anglo-Saxon scientists favored the theory that granite originated as basalt, while most western Europeans, especially the Germans (such as Helmut Winkler and Harold Mehnert), preferred a theory of sedimentary origin, or palingenesis.

After 1966, interest in granite's origin died out and researchers began to look into the problem of the origin of basalt. In spite of some famous debates, notably that between Mike O'Hara of Scotland and Dave Green and Ted Ringwood of Australia, the origin of basalt aroused fewer antagonisms. Basalt's source is clearly the upper level of the mantle; its volcanic origin is indisputable. It was simply a question of determining what material produced basalt upon melting, what makes up, in other words, the upper mantle. I will discuss this research in more detail later.

Suffice it to say here that the solution became possible only when autoclaves capable of withstanding pressures of 10,000 to 30,000 atmospheres became available and that today we know that fusing a mixture of 5 to 30 percent peridotites produces a basaltic melt comparable in composition to those found on the earth's surface.

In the course of a few years, experimental petrology was able to define the conditions in which the principal rocks of the earth's surface are formed. It is easy to see how such results would capture the scientific community's attention completely and how the old Wegenerian hypothesis would be totally neglected. Wegener's approach had brought the various fields within the earth sciences together; the development of numerous specialties worked against this ecumenism. One of these specialties, however, would re-awaken interest in the theory of continental mobility. Nothing predisposed the austere discipline of geomagnetism to such a role; the field was isolated from the mainstream of earth sciences. In fact, although the reemergence of mobility theories through magnetic studies may seem incongruous, it was to happen twice, as we shall see.

The Terrestrial Magnetic Field

Those who study geomagnetism are no doubt the least known and the most mysterious of all the earth scientists. Picking up signals that reveal activity in the core of the earth or the sun without knowing their exact cause, they are like cardiologists who listen to the beating of a heart without knowing what causes it. The sense of mystery about forces that act at a distance without being "visible" is so firmly set in our minds that we use the term *magnetism* to describe certain extraordinary qualities of human beings. Of course the use of this expression has reinforced the strangeness of the idea.

The mystery of magnetism is not a new one. The Chinese, who were less sensitive than we are to the notion of immediate cause and effect and much more so to the ideas of field and environment, discovered the magnetic field in the Han period and measured it precisely in the year 1040. In Western civilization this phenomenon became of interest only in the Renaissance. The work of William Gilbert and Karl Friedrich Gauss, and later of André Marie Ampère, James Clerk Maxwell, and their followers, illuminated its "primary cause," but the science of the earth's magnetic field remained impenetrable to Platonic reasoning. Before we discuss the causes of this, let us look at the central character in the play, the magnetic field.

As everyone knows, all magnetic bodies or magnets attract and are attracted by other magnets. The attraction is such that for each magnet in the form of a needle or a bar a north pole and a south pole can be defined. Like poles repel each other, and opposite poles attract. Electric currents are capable of acting like magnets; they attract or are attracted by other magnets.

Everyone also knows that a magnetized needle indicates, at least approximately, the north and south geographic poles. This is the principle of the compass, without which maritime shipping would have remained confined to the coastal trade. One can imagine that something must orient the needle, that a magnet in the earth's center must cause it to take the north-south direction. In fact, this is what William Gilbert, physician to Queen Elizabeth I, assumed at the beginning of the seventeenth century. With the help of some theoretical principles one can calculate the direction and intensity of the magnetic field at any point on the earth's surface. To a first approximation, the force lines of this field resemble those formed by iron filings around a bar magnet. Toward the end of the nineteenth century ingenious and often complex devices—magnetometers—were used to measure the terrestrial magnetic field. The quantitative exploration of the earth's magnetic field began then and expanded to include measurements taken over larger areas of the earth and over long periods of time.

Geographical exploration has given us the ability to draw the map of the earth's present-day magnetic field. This map is determined by three parameters: the orientation of the field (two angles), and its intensity. The magnetic field varies with latitude, location, and the composition of the underlying terrain. To describe geographic variations, magneticians defined *magnetic anomalies* representing the difference between the real, measured field and the theoretical field calculated according to Gilbert's hypothesis. If the measured field is stronger than the calculated one, there is a positive anomaly; if it is weaker, the anomaly is negative. Magnetic anomalies played a central role in the resurgence of Wegener's theories.

Observations taken over a period of time are more numerous and therefore more discriminating. Observatories have been tirelessly recording variations in the orientation and the intensity of the magnetic field at various points on the globe day and night for dozens of years; there are records of the magnetic field for Paris or London dating to 1838. An analysis of time series shows variations

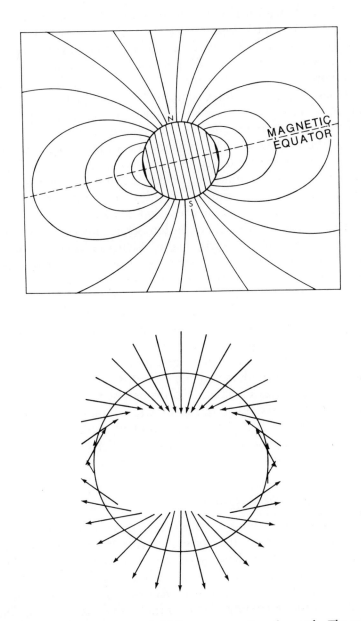

FIGURE 14   *Top:* Magnetic field lines surrounding the earth. The earth's magnetic field is like that created by a dipole magnet at the center of the earth. *Bottom:* The direction and intensity of the earth's magnetic field. Note that in the southern hemisphere the magnetic field is directed outward, whereas in the northern hemisphere the magnetic field is directed inward. The orientation of the magnetic field uniquely defines a latitude at the earth's surface.

whose frequency is itself variable: rapid variations, slow variations, and secular variations. Rapid variations can be observed daily; slow variations appear decades apart; secular variations, once a century.

Little by little the study of temporal and spatial data allowed researchers to identify the sources of the observed variations and anomalies. Rapid variations were found to originate in the high atmosphere of the earth, where ionized atoms interact with the flux of charged particles and electromagnetic waves coming from the sun (the solar wind). The movements of the atoms induce the magnetic variations measured in the ground, and hence the ulti-mate source of rapid variations is solar activity. The source of the slow and the secular variations is the core of the earth. Today everyone agrees to this double causation, but such was not always the case! After a detailed examination of the slow waves showed that their origin was the earth's core, scientists sought to under-stand the mechanism of their formation. These palpitations, these more or less regular tremors, are so rapid (in the order of years) compared with the length of geologic phenomena (in the order of millions of years) that they seem like the random swirls of a whirlpool. The earth's outer core is fluid and made of iron, which is an excellent conductor. The movement of a conductor in a magnetic field creates a current, and a current itself produces a magnetic field. To understand the way this works, however, we will have to investigate further. Unfortunately, the temporal series that we have are ridiculously short—a few hundred years of the 4.5-billion-year age of the earth, a factor of 20 million! One way to obtain longer temporal series is to determine the ancient magnetic fields fossilized in rocks, the subject I turn to next.

The Fossil Magnetic Field

In 1853 Macedonio Melloni, a Neapolitan refugee in Paris, noticed that volcanic lava had a distinct magnetism. In other words, each volcanic rock seemed to be a "magnet." Melloni hypothesized that this magnetism was acquired as the lava cooled and that its direction was that of the ambient magnetic field, the terrestrial field at the time of cooling. His hypothesis was elaborated in the experiments and observations of Bernard Brunhes in 1906 and of Paul L. Mercanton between 1910 and 1930. A theoretical explana-tion was furnished for it by another Frenchman, Louis Néel (who later won the Nobel Prize), following discussions with Émile Thellier, a devoted experimentalist.

This phenomenon is understandable for rocks that contain

magnetizable minerals (the obvious example being magnetite). When a magnetic field comes in contact with a magnetizable material, the material becomes magnetized; when the field is terminated, the material retains a remnant magnetism. The process works in an analogous way for volcanic rocks, which all contain a little magnetite. Above a certain temperature, called the Curie point, magnetite registers no magnetic field. In cooling, however, it registers a very small fraction of the magnetic field in which it crystallized. Thus volcanic rocks are memory banks of the earth's magnetic field.

Although the first studies of the earth's magnetic field were completed in 1936, it was only after the Second World War that this "magnetic memory" was used intensively. Émile and Odette Thellier and Alexandre Roche in France, Takesi Nagata in Japan, and Johann G. Koenigsberger in Germany were the patient and meticulous craftsmen of the method. All the researchers at that time had a common objective: to get long temporal series so that they could study the history of the magnetic field and determine its origin. The study of the fossil magnetic field accelerated swiftly in about 1950, because of an unexpected technological result of Patrick Blackett's unsuccessful research on the origin of the magnetic field.

In 1947 Blackett proposed the daring hypothesis that the magnetic fields of the planets, the stars, and every physical system are a consequence of the rotation of objects. Thus the terrestrial magnetic field must be produced by the earth's rotation. Blackett undertook an elaborate experiment on generating magnetic fields by rotation, which yielded two important results. The first was the failure of his hypothesis. With great personal courage Blackett, who was awarded the Nobel Prize in 1948, published his results in 1952 as a "negative experiment." The second result, which is what interests us here, is that Blackett invented an instrument that measures very weak magnetic fields: the astatic magnetometer.

This instrument and later refinements of it allowed researchers to measure the weak "fossil" magnetic fields in rocks. Around 1949 Blackett and Keith Runcorn, his assistant at the University of Manchester, decided to use the astatic magnetometer to measure the weak magnetism found in volcanic rocks and hired a student worker for the summer of 1950 for this purpose. The student, Ted Irving, collected enough results in one summer to convince the researchers of the importance of the experiment. But the team split up, Blackett going to Imperial College in London while Runcorn returned to Cambridge. Rather than terminating the enterprise, this separation marked the beginning of a competition between two respected laboratories.

The Revival of Continental Drift

In fact, the two English teams developed the new discipline of paleomagnetism with remarkable vigor. Early successes encouraged them to try to measure fossil magnetization in sedimentary rock. The fossilization of the magnetic field in sedimentary rock is acquired either through the orientation of magnetic particles as they are laid down in the sediments or through chemical transformation of soft sediment into rock, called diagenesis. The fossils they contain make sedimentary rocks easy to date, and if their age and the intensity of their magnetic fields can both be determined, a precise temporal series can be reconstructed.

Runcorn and Irving used an English Triassic sedimentary formation for their first systematic study. But the results were a surprise to them. They showed that the orientation of the fossil magnetic field for the English Triassic (200 M.Y.B.P.) did not coincide with the present-day orientation of the magnetic field. Could England have undergone a rotation over the course of geologic time?

The researchers at Imperial College decided to try to verify this audacious hypothesis in other countries, at first those in the British Empire: South Africa, Australia, Canada, India. In particular they studied the volcanic rocks of Deccan, India, laid down from the Triassic to the Tertiary periods in eruptions whose lava flows cover almost a third of the surface of India. Again, the results were surprising. Before the Tertiary (60 M.Y.B.P.) the orientation of the magnetic field was *external*—that is, it was directed toward the surface of the earth. Afterward the orientation varied continually until it became *horizontal* at the beginning of the Tertiary, and then gradually it became *internal*. Remembering the way the terrestrial magnetic field is located along a meridian (see Figure 14), and mindful also of Wegener's hypothesis of continental mobility, the researchers combined the two ideas and postulated that India had moved slowly northward between the Jurassic and the end of the Tertiary. Continental drift had reappeared! At first the Newcastle group to which Keith Runcorn had emigrated disputed the Londoners' mobilist interpretation. Runcorn noted that the conclusion that the continent had moved was based implicitly on an unverified hypothesis: that of the fixed orientation of the terrestrial magnetic field over the course of geologic time.

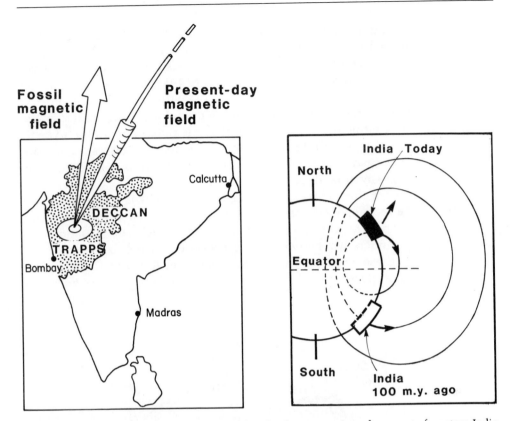

FIGURE 15 *Left:* The Deccan Trapps are basaltic lavas covering a large part of western India. This figure shows the direction of remnant magnetization of the Deccan lavas and the direction of the present-day magnetic field in India. *Right:* The discrepancy between the two is explained by the difference between where the Deccan Trapps were formed and where they are now. The lavas were laid down 60–100 million years ago, when India was in the southern hemisphere, and were magnetized by the earth's magnetic field characteristic of that time and place. India has since drifted northward, to a latitude where, coincidently, the earth's magnetic field has an orientation almost opposite to that of the latitude at which the lavas were formed.

What if the poles had been drifting too? Wegener himself had suggested that possibility.

Runcorn therefore proposed to chart the position of a paleomagnetic pole for several time periods and for a variety of areas and then to compare the results before coming to a conclusion. The first step in this task was to define a curve of "polar wandering" for Europe over the course of geologic time. After Europe the Newcastle group studied America. On the basis of these studies, they were able to define another curve of polar migration for America distinct from that charted for Europe and merged with it only before the Cretaceous period. The apparent divergence of the two curves disappeared if the Atlantic Ocean was left out before the Cretaceous. At that point Runcorn and his students, converted to the theory of continental drift, proposed a method for reconstructing successive global geographies: move the continents in such a way that the polar-migration curves fit together for all of them in all periods. After some debate "Newcastle" accepted "London's" interpretations by taking the long way around: polar wandering!

Since these ideas were worked out paleomagneticians have used the technique of plotting "polar paths" to reconstruct the successive positions of the continents: all that is necessary is to move the continents two by two until the pole for a given period coincides for those continents under consideration. Kenneth Creer, Irving, and Runcorn used this technique to chart the successive positions

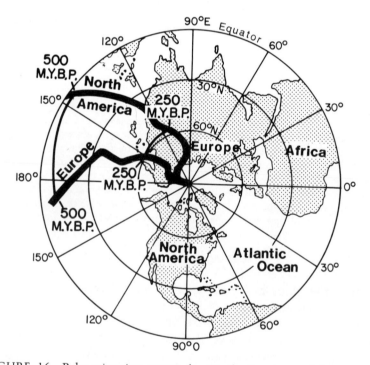

FIGURE 16 Polar-migration curves for North America and Europe. To overlay these two curves we would have to move North America eastward, which would close the Atlantic Ocean and by uniting North America and Europe form the ancient continent of Pangaea. [Figure from Keith Runcorn.]

of the continents in the various geological epochs. Their reconstructions resembled Wegener's to an astonishing degree, with a few exceptions, the most notable being that of India. As the reader may recall, Wegener joined India to Laurasia. He thought that when Pangaea split into pieces, India had drifted northward with the Asian continent. This idea had led Argand to postulate an intra-Asian contraction to explain the birth of the Himalaya. The paleomagneticians, on the other hand, associated India with Africa, Australia, and Antarctica, making it part of Gondwanaland. India detached from Gondwanaland and migrated northward only about 130 million years ago, in the Cretaceous period. Thus, the Himalayan mountains appeared to be the result of India's *collision* with Eurasia, not of an intracontinental folding process. Paleomagnetic reconstruction therefore contributed an important new idea, different from those of Wegener and Argand. I will examine some of the consequences of this idea later.

But the paleomagneticians did not stop at 200 M.Y.B.P. Wegener had imagined that the single continent, Pangaea, had existed since the beginning of time and that it had not started to break apart until the Triassic (220 M.Y.B.P.). For him, continental drift was therefore unidirectional and irreversible: a single continent broke into pieces that continued to fragment over time. Because paleomagnetic methods do not depend on geomorphologic reconstructions, they allow us to study continental drift dating to more than 200 million years ago, all the way back to the Paleozoic era. This task is not without difficulties: for one thing, the older the materials one works with, the greater the chance that their initial magnetism has been masked by secondary signals. There are also stratigraphic and chronologic problems: to reconstruct the past by paleomagnetic methods, each sample must be dated very precisely. The *absolute* dating of Paleozoic and Precambrian rocks is much more uncertain than that of recent rocks. An analytical error of 5 percent results in an error of 5 million years for a sample from the Cretaceous period (100 M.Y.B.P.), but the same margin will result in an error of 100 million years for a sample 2 billion years of age. Nevertheless, paleomagneticians persevered in their task, while admitting uncertainties in their reconstructions for the Paleozoic. They rapidly affirmed the existence of continental drifts that had taken place more than 300 million years ago. Wegener's ideas had been outdated. Continental breakups, drifts, and collisions and the births of supercontinents must be considered simply so many successive phenomena that do not conform to a unique and irreversible

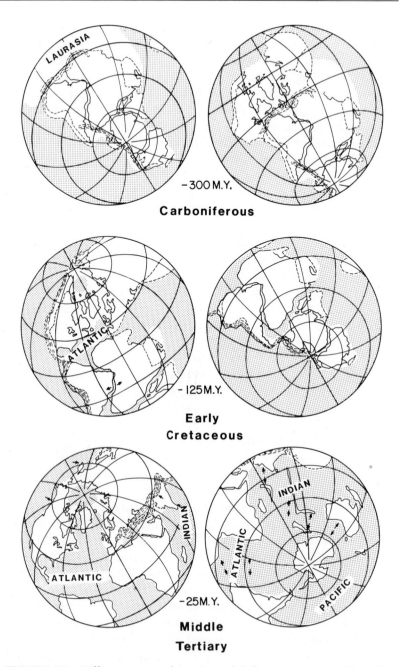

FIGURE 17 Different stages of continental drift as reconstructed by paleo-magneticians around 1958–1960. Continental locations are shown for the Carboniferous, the Cretaceous, and the Tertiary periods. It is interesting to compare this reconstruction of continental drift with Wegener's (see Figure 4).

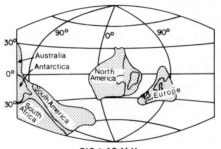

−510 ± 40 M.Y.

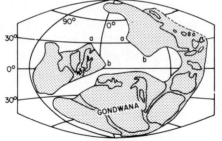

−380 ± 30 M.Y.

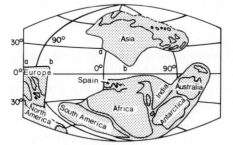

−340 ± 30 M.Y.

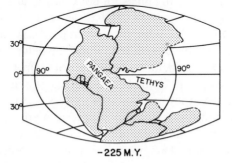

−225 M.Y.

FIGURE 18 Reconstruction of the positions of the continents before the formation of Pangaea. (Based on the paleomagnetic measurements of J. C. Briden and his collaborators.)

progression. The earth is not evolving from a single continent toward a collection of ever smaller and more numerous continents; in fact, earth's history shows an anarchic alternation between periods in which the continents gather together and periods in which they split apart. I will return to this prophetic view in the last chapter.

These beautiful reconstructions, however, were greeted with skepticism from the rest of the scientific community in the 1950s: "The new ideas call into question the paleomagnetic method itself." "Was the proper statistical method used?" "Is it legitimate to extrapolate the hypothesis that the earth's magnetic field is bipolar to the distant past?" Harold Jeffreys, an unyielding opponent of all mobility theories, even claimed that the hammer used for taking rock samples was responsible for their magnetism. The English scientific community, which had been well prepared by Arthur Holmes's writings, should have been receptive to the idea of continental drift, but Keith Runcorn and his colleagues were greeted with skepticism in the United Kingdom, as well. In the United States not only skepticism, but open hostility, was publicly expressed. With the exception of some Dutch paleomagneticians, European scientists gave drift theory a polite but icy welcome. It was generally thought that this theory could be accepted only by specialists. Thus, confined within their specialty and regarded with suspicion, paleomagneticians had neither the courage nor the opportunity to reopen the debate over continental drift with earth scientists in other fields.

Inversions of the Magnetic Field

If Keith Runcorn's stirring pleas for continental drift hardly moved the Americans, who were as "fixist" as their Soviet colleagues, they at least awoke a growing interest in paleomagnetic science in the United States. The Americans, whose taste for figures led them to apply physical methods to geology, took up the study of the earth's fossil magnetic field. Prudently leaving aside the controversial drift theory, they attacked the no less controversial hypothesis of reversals of the terrestrial magnetic field.

In 1906, when Bernard Brunhes discovered that the magnetic direction of volcanic lava samples from the mountains of the Auvergne in France was opposite to that of the present-day magnetic field, he also noticed that clay soils baked by direct contact with the hot lava acquired the same orientation as the underlying lava—normal orientation (defined as the orientation of the present field) when the lava's magnetism is normal, or inverse orientation (opposite to the current orientation) when the lava's orientation is inverse. Using laboratory experiments, he showed that bricks (baked clays) acquire the orientation of the magnetic field in which they are cooled. In 1926 Mercanton confirmed Brunhes's observations and suggested the possibility of reversals of the magnetic field in the past. The north-south dipole could have had a south-north orientation in other periods! This grandiose hypothesis awakened little interest at the time.

In the same period Motonori Matuyama of Japan shed light on this phenomenon by adding a temporal dimension to it. He noticed that some lava flows from Japanese volcanoes showed positive magnetic anomalies and that other flows showed negative anomalies. He measured the magnetism in rocks from various flows and ascertained that negative anomalies correspond to inverse magnetism, thus confirming Brunhes's and Mercanton's observations. According to Matuyama, the terrestrial magnetic field first reversed itself and then became normal again. This idea too, which should have excited the geologists of the time, was greeted with indifference and fell into oblivion.

In 1950 John W. Graham, then at the Carnegie Institution of Washington, became interested in the phenomenon and proposed a new explanation for inversions that attributed them to a phenomenon of solid-state physics. Under some circumstances, the presence of a magnetic field during the cooling of certain minerals causes them to take on a field opposite in direction to that of the background field. Louis Néel furnished a theoretical base for this explanation, and Nagata and Seiya Uyeda demonstrated experimentally the existence of the phenomenon in Japanese lava. The hypothesis of autoinversion is exactly the opposite of that of reversals of the earth's magnetic field. A vigorous and contradictory debate over this idea put paleomagnetism back in the spotlight. Is magnetic memory sometimes recorded as the opposite of a field whose direction is constant, or is the field itself subject to periodic reversals?

This question was first answered through the efforts of researchers at the University of California, Berkeley, two of whom played especially decisive roles. By 1960 John Reynolds, of the physics department, had improved the method of dating radioactive potassium, making it possible to measure very small quantities of argon, the decay product of potassium 40, and to date young rocks (less than 10 million years old) with reasonable precision. John Verhoogen, of the geology department, took up the problem of autoinversions and interested his students in this work. The two groups cooperated on obtaining precise ages and clearly defined magnetic orientations for the same basaltic samples, and interesting results came pouring out. Allan Cox and Richard Doell rapidly noticed that autoinversion did not explain their findings for lava flows from Idaho. They reconsidered the idea that the earth's field had reversed, as put forward by Mercanton and Martin Rutten of Holland, and decided to test it by applying the method of absolute chronology. If all lavas with inverse magnetism were the same age, whatever their mineralogical features or geographical position, that would mean that the terrestrial magnetic field could reverse.

By the 1960s such a possibility appeared less improbable, because knowledge about the origin of the magnetic field had increased. Walter Elsasser of Princeton and Teddy Bullard of Cambridge, England, proposed the model of a dynamo situated in the earth's core; the unstable behavior of this dynamo could explain "periodic" reversals of the magnetic field it created. This idea reinforced the views of those who supported the reversal theory.

The existence of reversals in the magnetic field was "proved" between 1960 and 1966 through the patient work of the two Berkeley groups: Allan Cox, Richard Doell, and Brent Dalrymple worked for the U.S. Geological Survey in Menlo Park, south of San Francisco; the Berkeley-related scientists Ian McDougall and François Chamalaun, of the Australian National University, operated in the Pacific and the Indian Ocean. From recent volcanic lavas the researchers established a table of inversions for the last 4 million years, a table that applies not only to the United States but also to Europe, the Pacific, and Australia and that is valid worldwide. This table shows shifts in polarity that vary in both duration and frequency. During *magnetic epochs* the terrestrial magnetic field keeps the same direction for several hundred million years, whereas *magnetic events* last for much shorter periods within epochs.

Paleomagneticians named the epochs after the great pioneers of terrestrial magnetism. The epoch that extends from the present to 600,000 years ago is called the *Brunhes epoch*, and the preceding one, from 600,000 to 2.4 million years ago, is the *Matuyama reverse*. The events were given the names of the sites from which the polarities were identified, such as Jaramillo (95 M.Y.B.P.), Gilsa (1.6 M.Y.B.P.), Olduvai (1.9 M.Y.B.P.), and so on. Thus was established a "magnetic scale" with a very characteristic signature thanks to its series of epochs intersected by events.

Neil Opdyke and his colleagues at Lamont Geological Observa-

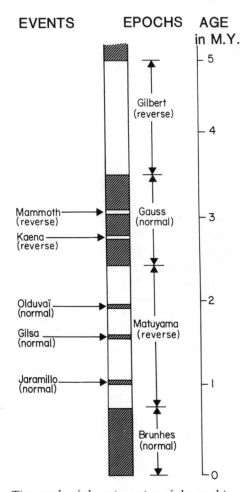

FIGURE 19 Time scale of the orientation of the earth's magnetic field. We distinguish between magnetic *epochs* of long duration and *events* of short duration. By standard convention dark bands represent periods when the orientation of earth's magnetic field is normal (same as the present), and white bands indicate periods when the orientation is reversed.

tory confirmed the existence of the inversions through an ingenious experiment. Instead of using volcanic lavas as samples, Opdyke studied marine sediments. As we have already seen, sediments do record a magnetic field, but taking samples of them presents a problem. Samples are taken by forcing a hollow cylindrical piston down through the soft sediments. The cores brought back to the surface in this way look like big sausages cut in slices, each slice a sample of a different sedimentary layer. Unfortunately, since it is difficult to determine the spatial orientation of the cores, it is also difficult to determine their geographical orientation. If it is possible to measure the magnetic orientation of the various layers, how can it be related to the local orientation? Opdyke used sediments from the Antarctic Ocean, where the magnetic field is almost vertical and therefore slanted very little in comparison with the core. Inversions in these cores were easy to read. Reversals were found to exist in these soft sediments for which the mechanism of autoinversion is not possible. Better yet, by using micropaleontology to date samples, Opdyke determined that the time scale of inversions recorded in the sediments coincided with that established for basalts. This chart of inversions showed an unstable magnetic field, one whose orientation reversed for periods of variable length with no clear regularity.

In other words, on a scale of millions of years or a day, the magnetic field seemed unstable, changeable, variable—so irregular that the temporal message embedded in rocks clearly indicates the period in which it was recorded. The rebirth of the theory of continental drift would begin with this magnetic message, a magic key to unlock the history of the oceans.

The modern world has a tendency to consider scientific research *the* rational activity par excellence. Those who think that it is possible to plan research and predict its results would do well to think about these studies of terrestrial magnetic fields, which never directly answered the question posed at the outset. They always led to a fundamental answer in an unexpected area. So goes science, defying plans, ideologies, and ideologues, but also the most rigorous scientists.

From Wegener's death in 1930 to the resurrection of his theory in 1961, the earth sciences underwent an extraordinary development. The mapping of the continents and oceans gave us for the first time an integrated picture of the earth. The least hospitable and most inaccessible regions of the globe were no longer mysterious. The structure of the earth's interior, which had been the object of research by geophysicists and astronomers for centuries, began to give up its secrets. A solid, comprehensive body of information (which future studies would largely confirm) was becoming available both for the directly observable skin of the earth, and for the less accessible regions—the ocean floor and the interior of the earth. The formation of materials from the depths, requiring high temperatures and pressures, was explained theoretically as well as re-created in the laboratory. The science of rocks made a spectacular and decisive leap in a short time, and the study of the magnetic field was advancing in giant steps also. Scientists were able to separate the field whose origin is the sun from the field whose source is the fluid imprisoned in the exterior core; they were able to measure magnetic fields and their fluctuations and to construct models that explained their observations. Moreover, in this period the extraordinary science of paleomagnetism developed. Sedimentary and volcanic rocks were found to be magnetic memory banks; they record the directions of ancient magnetic fields. Accumulating data in these fields revealed two unexpected "new" phenomena: the movement of the continents and the inversions of the field.

In the end, however, the new findings retained the interest of specialists only; they did not penetrate the scientific community as a whole. In spite of the abundant harvest of information, scientists still tended to see their own specialties as autonomous entities. To nonspecialists the earth sciences appeared a mosaic of separate disciplines connected by some common areas, but each having its own central interest. Exclusive specialization was probably necessary, for the problems to be solved were so difficult, diverse, and important, but it obscured the idea that the central object of study was the whole earth, that the various aspects of the globe possessed an underlying coherence. In the end the feeling that a geophysicist and a paleontologist or a geochemist and a structural geologist had nothing in common was the barrier that proved the most difficult to cross when drift theories were reborn.

To understand the mental progression toward widespread acceptance of the theory of continental drift, we must be aware of the proliferation of advances in the earth sciences that preceded it. The natural sciences pass through building phases in which profound organizing principles are strongly felt and explosive phases in which the centrifugal force of a multitude of specialties causes common goals to fade from view. Alternations between phases of expansion and those of consolidation are a necessary condition of scientific progress. In fact, how could the establishment of the geologic world map or the structure of the earth *not* be considered an end in itself? How could the development of the geologic clock *not* occupy careful experimenters completely? How could the bounds and rebounds of magnetic science *not* be completely enthralling? By disdaining Wegener's theory earth scientists had lost their common denominator but not the range of their interests. The spectacular development of their separate fields no doubt explains why drift theories were not restored to a place of honor

SEAFLOOR SPREADING

3

AROUND 1960 the theory of seafloor spreading sparked a revival of interest in the idea of continental drift. I will examine the genesis and development of this theory as it is recalled in the collective consciousness of science. After that I will return to some parts of the story, because when a scientific theory achieves the status of a well-established principle its adherents tend to invest it with a rationality that hides the quirks so evident at its beginning.

The Theory of Seafloor Spreading

To a book published by the Geological Society of America in honor of A. F. Buddington in 1962 Harry H. Hess contributed an article entitled "The History of Ocean Basins." This article constituted the debut of the idea that the ocean floor is spreading. Hess, a professor of geology at Princeton, had served in the navy in World War II, rising to the rank of captain of the *Cape Johnson*. The *Cape Johnson* sailed the South Pacific and, during the course of military missions, drew the bathymetric map of the regions it crossed. During these missions Hess became familiar with the morphology of the ocean floor with its ridges, trenches, and guyots (seamounts), and he began to wonder about the origins of these features. He was particularly interested in guyots, submarine volcanic formations with flat tops, and in 1948 he proposed a hypothesis for their origin. Some will see early germs of the idea of *seafloor spreading* in his proposal. Hess said that guyots begin as aerial volcanoes on oceanic ridges. Wave erosion truncates the volcanic cones horizontally at their summits, and then the flat-topped mounds "slide" slowly down the flanks of the ridges and disappear into the ocean. Hess believed that this process took a long time, and that the Pacific guyots were probably Precambrian, that is, several billion years old. The discovery of fossils from the Cretaceous (130 M.Y.B.P.) at the bases of guyots in the 1950s destroyed his time scale and also his theory, but Hess did not lose interest in the ocean or the rocks that were being dredged up from its floor. In particular he studied the problem of serpentine, a shimmering green mineral that is a product of the interaction of water with peridotite. I mention these circumstances to explain how Hess altered his theory by 1962 in response to them.

Hess postulated that the earth's mantle is traversed by wide *convection currents.* The surface manifestation of the rising portions of these currents is the mid-oceanic ridges; the descending parts are found in the great trenches that border the oceans. The ocean floor is formed at the ridges, drifts systematically away from them on both sides, and then plunges into the mantle at the trenches. Thus the seafloor is a *conveyor belt* for material that is continuously recycled in the mantle. Using the first data from marine stratigraphy, Hess calculated a speed of one centimeter per year for the drift of the ocean floor (remember that Wegener envisioned speeds ten times as fast!). Continuous spreading explained the considerable altitudes of the oceanic ridges (as detected by Bruce Heezen and Maurice Ewing), the enormous quantities of heat emitted there (as measured by Teddy Bullard, Arthur Maxwell, and Roger Revelle of Scripps starting in 1956), and the "migration" of the guyots, all at the same time.

If seafloor spreading does occur, the zone in which it is produced must be dilated or distended by the flow of hot material coming from beneath. If hot material is rising toward the surface

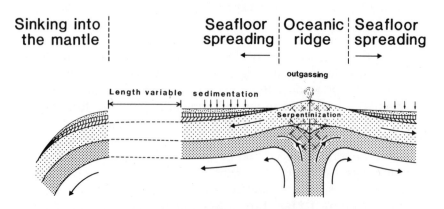

FIGURE 20   Harry Hess's model of seafloor spreading. Hess thought that at a mid-oceanic ridge water expelled from the mantle reacts with igneous material and serpentinizes the oceanic crust. As the seafloor moves away from the mid-oceanic ridge, sediments are deposited with time. Finally the seafloor plunges into the mantle at oceanic trenches.

to form the ocean floor, it must transport heat with it. Therefore, the flux of heat at the ridges must be in excess of the average terrestrial heat flux. Hess said that as the continents are pulled passively along on the conveyor belt, their low density prevents them from being drawn down into the mantle, so they always remain at the surface, forming the "unsinkable memory of the earth's history." On that point Hess agreed with the theory of the permanence of oceans and continents dear to the hearts of old-time geologists.

Hess's article seems profound and prophetic to us today, but at the time of its publication scientists did not see it that way. Even its author considered it a daring attempt and called it a "geopoetical essay" as if to emphasize its lack of rigor. His object was to explain two oceanographic observations that were difficult to account for in theories then current: on the one hand the absence of very old oceanic rocks (ocean dredging had never brought up rocks older than the Jurassic) and, on the other hand, the thinness of the sedimentary rock layers on the ocean floor. Millions of years of continental erosion had certainly produced colossal quantities of sediments. Where had they gone? Hess's theory provided a single answer to both questions. Old oceanic rocks, like old sediments, did not exist on the ocean floor because they were pulled down into the mantle through the trenches and therefore disappeared from the terrestrial surface.

Hess's caution in proposing his theory was also reflected in his choice of publication outlet. Rather than submit the article to a widely circulated journal, he published it in a book honoring a friend, a festschrift for which editorial standards could be expected to be less strict. At the same time he circulated a great number of prepublication copies to test the reactions of American geologists. They were so apathetic and reserved that Robert Dietz had time to publish an article in which he developed analogous ideas before Hess's contribution appeared. Later Dietz admitted good-naturedly that he had known about Hess's manuscript and that his own did not have precedence. But his honesty would go unrewarded, for his name gradually disappeared from the list of originators of the spreading theory even though his apparently analogous article turned out to be very different from Hess's.

Dietz coined the expression *seafloor spreading;* moreover, he also realized that the mechanical entity that drifted laterally was not the oceanic crust but a much thicker layer that he called the lithosphere. This idea was the germ of the idea of plate tectonics. But let's not run ahead of the story, because the desire for simplification, the wish to identify the unique author of a discovery, always distorts reality, which is much more complex. In the present case, let me emphasize that it would be unfair to forget the name of Robert Dietz.

Be that as it may, the theory developed by Hess and Dietz encountered nothing but skepticism. Only J. Tuzo Wilson took it up, adding a new argument in favor of it. He maintained that the age of oceanic islands grows with their distance from oceanic ridges: the farther away they are, the older they are. This proves that they were formed at the ridge and were carried away from it by the oceanic conveyor belt. Paradoxically, this support only reinforced the general skepticism. Wilson had been the unquestioning champion of the theory of thermal contraction, finding a multitude of structural arguments in favor of it, and then the intransigent

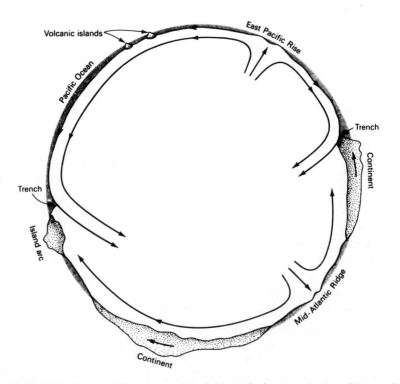

FIGURE 21 Equatorial cross-section of the earth showing the distribution of spreading ridges and subduction zones. Note how the spreading ridges and converging trenches do not alternate in a regular fashion but instead are haphazardly distributed on the face of the earth.

defender of S. Warren Carey's theory that the earth had become larger over time and that the continents were separated from one another as it expanded. The scientific community tends to distrust anyone who changes his opinions too often or who appears to be attracted by novelty above all else. In this case its distrust was reinforced by similarities between Hess and Wilson. Like Wilson, Hess had often entertained daring theories (such as his theory of guyots). His ideas on the genesis of mountain chains were no less bold: he denied the existence of "ophiolitic associations," which I will discuss later on.

The agreement between these two scientific adventurers did nothing to reassure the conservative scientific community. After years of wandering in the wilderness, however, the two had finally stumbled upon a promising path and were moving in the right direction.

The Zebra Skin

The study of the earth's magnetic field, which had been the source of two unexpected discoveries, produced a third revelation in the undersea world.

After the war, when modern oceanography developed momentum, researchers began mapping the magnetic fields of the ocean floors. Oceanographers took up the practice of trailing magnetometers behind ships (to place them as far as possible from the magnetic, metallic, and electrical disturbances caused by the ships themselves) and recording the magnetic field continuously. Thousands of magnetic measurements were then mapped. Actually, the values of the magnetic field itself were not recorded, but the anomalies that express the difference in intensity between the measured field and the theoretical field. The advantage of using anomalies was that it eliminated the dependence on latitude of the magnetic field, so that various regions of the globe could be compared.

The resulting maps of the marine magnetic anomalies remained mysterious for a long time. They showed no regularities, no links between magnetic anomalies and other measurements such as depth or distance from shore. Around 1955, however, when enough data had been accumulated, regularities did begin to emerge from these pictures.

Victor Vacquier and Bill Menard of Scripps observed that when the great fracture zones of the eastern Pacific (Clipperton, Sequiero, Mendocino) are crossed, the magnetic anomaly disappears.

Detailed mapping of these zones showed that the anomalies reappear on the other side of the fault, but they are displaced (or offset). Vacquier's group (notably Ronald Mason) demonstrated a much more curious fact: the existence of bands of positive anomalies (bands in which the measured field is stronger than the theoretical field) and bands of negative anomalies. On some areas of the ocean floor these bands alternate regularly. If the positive anomalies are colored black and the negative anomalies white, the map looks like a zebra skin. Around fracture zones, the "zebra skin" is torn and its pattern shifted.

Then another new fact was discovered. Near an oceanic ridge—the East Pacific Rise or the Mid-Atlantic Ridge—the bands lie parallel to the axis of the ridge. Not only are they parallel, but they are symmetrical to the ridge axis. If the axis is a black band, the two bands adjacent to it are white, the next ones black, and so on. How could this strange picture be interpreted?

Lawrence Morley of Canada and Fred Vine and Drummond Matthews of England had the idea of combining three pieces of information that seemed unrelated at first glance: Motonori Matuyama's interpretation of the magnetic anomalies of volcanic flows; Allan Cox's and his colleagues' chart of the reversals of the terrestrial magnetic field; and Harry Hess's theory of seafloor spreading.

The strength of a magnetic field measured at a given point is the sum of the global field, created by the internal dipole, and the local magnetic fields whose source is near the surface. It is natural to suppose, therefore, that the magnetic field measured at sea results from the superposition of the global field and the field created locally by the ocean floor. The floor is made of basalt, which, as we have seen, fossilizes the magnetic field in existence at the time of its crystallization: basalts created in the past 600,000 years (the Brunhes epoch) show a normal field of orientation; those that crystallized between 600,000 and 2.4 million years ago (the Matuyama epoch) recorded a reverse magnetism, and so forth.

According to Hess, the fundamental function of the ridges was to manufacture the basaltic flows that immediately start to drift away from them on the conveyor belts on both sides of the ridge. The age of the basalts increases, then, the farther they are carried from the ridges. Remembering the magnetization of basalt and the existence of reversals, we can predict that we will find basalts of normal magnetization (those that flowed out during the Brunhes epoch) near the ridges and then, further away, basalts with reverse magnetism (those created during the Matuyama); beyond these, we

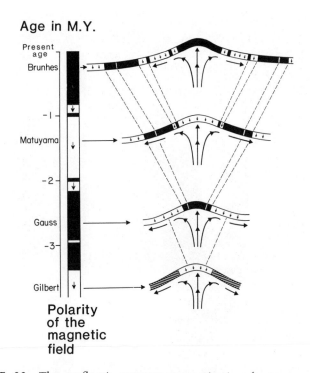

Age in M.Y.

Present age

Brunhes

−1

Matuyama

−2

Gauss

−3

Gilbert

Polarity of the magnetic field

FIGURE 22 The seafloor's remnant magnetization forms a pattern of alternating normal and reversed magnetic bands that record the magnetic field at the time the oceanic crust was formed. Because oceanic crust is formed within a narrow band at the ridge axis and then spreads away from the ridge with time, the alternation between normal and reversed magnetic orientation forms an accurate horizontal record of the magnetic time scale.

will find basalts of normal magnetization; and so forth. According to the principle of addition of magnetic fields, the small fields created by the magnetization of the basalts would be added to the present global magnetic field. Thus, the measured magnetic field *at* a ridge zone today would be slightly augmented by the fossil field of the basalts; if, on the other hand, measurements were taken far enough from the ridge to be on the part of the conveyor belt that was formed during the Matuyama epoch, the field of the basalts that is added to the present field would be negative, so the resulting field would be a bit weaker. The calculation of magnetic anomalies entails subtracting the theoretical global field from the measured one; the subtraction produces a positive magnetic anomaly in the first case, at the ridge zone (conventionally mapped in black), and

Fred Vine and Drummond Matthews

a negative anomaly in the second case, away from the ridge zone (mapped in white). This phenomenon is more complicated near the magnetic equator, however, because of the arrangement of the magnetic force lines there.

According to Morley, and also to Vine and Matthews, the zebra skin is the horizontal "projection" on the seafloor of the chart of reversals of the terrestrial magnetic field. The continuous creation of basalt at the ridges and the symmetrical drift on the conveyor belt provide a continuous record of the fluctuations of the magnetic field. The ocean floor is a memory bank for the terrestrial magnetic field. Thus, the interpretation put forward by Vine, Matthews, and Morley substantiates Hess's and Dietz's theory of the spreading of the ocean floor.

I call this interpretation the Vine-Matthews-Morley hypothesis, but in most texts on the subject (except Peter Wyllie's *The Way the Earth Works* of 1976) it is called the Vine-Matthews hypothesis. Vine and Matthews received many scientific prizes, but Morley never did. Is scientific memory selective?

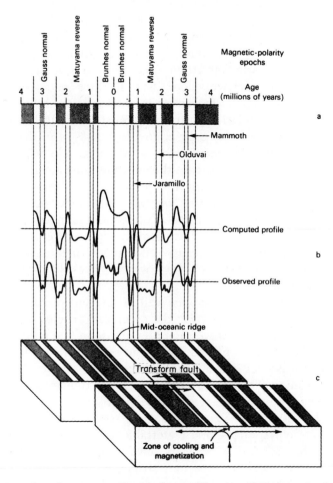

FIGURE 23 The two graphs in the middle are taken from Fred Vine's original figure of oceanic magnetic anomalies. The mean magnetic field intensity is subtracted from the field intensity measured locally and the resulting magnetic anomaly is plotted. The magnetic field due to alternating blocks of normal and reversed oceanic crust can also be calculated. The top shows the magnetic time scale, the bottom the pattern of reversed and normal orientation of the oceanic crust that produces the observed anomalies. The observed anomalies may be complicated by the presence of transform faults.

To understand such an oversight, we must look at that fundamental fact of scientific life: publication. When a scientist has completed a piece of original work, it is his duty to make it known—one could almost say that this is what he is paid for. So he writes an article and submits it to a specialized journal, which has an editor-in-chief and reviewers. The role of the reviewers is to decide whether an article can be published as it is, must be revised before publication, or should be rejected. The editor-in-chief chooses reviewers from among the scientists competent in a given area, collects their opinions, synthesizes them, and decides whether or not to print the article. The decision is not arbitrary; it is based on the reviewers' opinions. If three reviewers react negatively to an article, the editor-in-chief has no choice but to reject it. If their opinion is positive, he must accept it. His role as arbitrator comes into play only in doubtful cases or in the important instance when the person who wrote the article disputes the reviewers' opinions. In this case the editor obtains a reevaluation. The qualification of the editor-in-chief to carry out this task often depends on the level of the journal. The great international journals have renowned scientists as editors; more modest journals have less well known ones. Furthermore, the reviewers are members of the scientific community chosen for their competence; they are the author's "peers"—from which comes the name of the system, the peer review system.

This system works extremely well 90 percent of the time. It serves as a filter, helps to raise the quality of articles by forcing authors to clarify or improve their presentations, and prevents the financial waste that would result from an uncontrolled publication system—a waste of printing and money, but also of hours of fruitless reading. Thus, the acceptance of an article by a very good journal is a kind of index of scientific quality.

But like most filtering systems, peer review causes a certain number of delays, errors, and regrettable accidents. It sometimes happens that all three reviewers have a simultaneous lapse of judgment and reject an important article; more rarely it happens that unscrupulous reviewers delay an article so as to be able to publish their own results on the same subject first. Unfortunately, when the peer review system is examined critically, this minority of accidents is always mentioned, and the fact that the system generally works well is sometimes forgotten.

The "Morley case" is an example of such an accident. At the beginning of 1963, Morley submitted an article setting forth his theory to the British journal *Nature*. The article was refused on the advice of qualified reviewers. When he submitted it to the American *Journal of Geophysical Research*, he received another refusal. Fred Vine and Drummond Matthews submitted their article to *Nature* toward the middle of 1963. It was accepted immediately and published at the end of 1963. Then Morley's article was accepted for publication with Andre Larochelle as coauthor, but it did not appear until 1964 and credit for the discovery went to Vine and Matthews.

The refusal of Morley's article is understandable, though not excusable, if one remembers that at the time his interpretation of the zebra skin was considered unfounded by the *majority* of researchers. In the opinion of most of the scientific community the acceptance of Vine and Matthews's paper was the amazing thing. In fact, the Vine-Matthews-Morley theory would not be endorsed by the scientific world for several years.

At first opposition was voiced by marine magneticians and by the powerful groups of oceanographers at the Lamont Geological Observatory and the Scripps Institution. Their arguments fell into two categories. Many scientists who had worked on the magnetic mapping of the oceans and who knew the great generalizations and the complexity involved in it believed the zebra skin to be a rare phenomenon. In the Atlantic and the central Pacific, for example, magnetic anomalies do not make such simple patterns. The group at Lamont, following the lead of Maurice Ewing, tried to solve the mysteries of the ridges in the Atlantic. A detailed examination of the Mid-Atlantic Ridge brought to light a much more fundamental fact than the zebra skin: the central anomalies looked different from those on the outer flanks of the ridges. The former were clear, with neatly arranged fluctuations, whereas the latter became weaker the farther they were from the ridge and their fluctuations occurred farther and farther apart. In a 1966 article, Jim Heirtzler, Walter Pitman, Manik Talwani, and Xavier Le Pichon synthesized the information on the marine magnetic anomalies and attacked the Vine-Matthews-Morley hypothesis, without suggesting an alternative one. In 1967, concluding the thesis he had written at Lamont but was defending at Strasbourg, Le Pichon asserted that his work showed clearly that the theory of seafloor spreading was false! When participating in meetings or scientific seminars, the Scripps researchers were no less skeptical. The rest of the geophysical and geological community questioned the validity of the data used by Vine, Matthews, and Morley. Magnetic anomalies are usable only after they have been sorted out by a preliminary treatment whose legitimacy is not clear to the nonspecialist, who simply notices that the raw data that come out of the magnetometer in graphic form look extremely complicated. The symmetry of these recordings in comparison to the ridge is unclear, whereas the magnetic anomalies computed as the result of a variety of corrections are clear, simple, and symmetrical. To many people these corrective treatments seemed like doubtful "manipulations of the data," although they were, in fact, indispensable and perfectly legitimate.

Transform Faults and Spreading Speeds

Marine geophysicists were converted very quickly, however, by J. Tuzo Wilson and by Fred Vine himself. The two had just completed a stay at Princeton during which they talked at length together and with Harry Hess. In 1965 an article by Wilson appeared in *Nature* that developed the concept of *transform faults.* These fracture zones on the ocean floor can be a thousand kilometers long, are marked by morphological irregularities, and can displace the linear patterns of magnetic anomalies. The Vine-Matthews-Morley hypothesis led to the idea of large horizontal shifts in these faults, displacements on the order of thousands of kilometers. The faults appeared to bear witness to the movement of pieces of the ocean bottom, but how could they be integrated into the concept of seafloor spreading?

Wilson proposed that the faults be considered dynamic, not static structures. Take the example of a ridge offset by a transform fault. In the area between the two ridges, the two edges of the fault are continuously slipping past each other. Two fixed observers on opposite sides of the fault at a given point would move away from each other over an indefinite period. On the other hand, in those regions *beyond* the area defined by the two sections of ridge, the offset remains constant over time. Two observers situated on the same side of the ridge but on opposite sides of the fault at the same moment would travel along together throughout the spreading episode. This new class of fault had the amazing property of being active on one section of its length while showing a constant displacement over the rest of it. Thus, for example, the mid-oceanic ridges are cut into a series of discontinuous segments linked together by fracture zones. Since these faults are the borders between two conveyor belts, the displacement of the faults is equal to that which separates the two ridge sections. In contrast, a classic fault is a tear or cleft in the earth and therefore has a maximum displacement that dies out and vanishes at its ends.

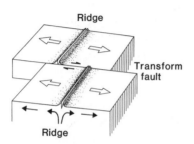

Ridge

Transform
fault

Ridge

FIGURE 24　A transform fault offsetting two ridge segments. In the area *between* the two ridge axes the plates on both sides of the transform fault move in opposite directions. In contrast, on the *far sides* of the ridge axes the inactive transform faults, or fracture zones, show no slip motion.

Wilson did not stop at ridge-ridge faults. He showed that the same mechanism displaces the trenches into which the oceanic crust plunges. These, then, would be trench-trench faults. The transform fault is like a relay that *transforms* a section of a structure that creates or destroys surface into another section of a structure that creates or destroys surface. From this point the obvious next step is to "cross over" and imagine faults linking ridges and trenches. The Denali Fault, for example, joins the Aleutian Trench (off the coast of Alaska) and the Juan de Fuca Ridge (off the coast of California).

This concept, an important one in the development of the new theory, is mentioned here simply to show that the structure of ridges, particularly that of the Mid-Atlantic Ridge, is very complex, with numerous offsets, and that as a result the zebra stripes are clear only if the magnetic surveys are sufficiently detailed. It is understandable that the results of these surveys did not always come out looking like zebra skins. Thanks to the idea of transform faults, Wilson was able to refute the objections of the Lamont group.

The Vine-Matthews-Morley hypothesis gained further support from another article published in 1965, this one coauthored by Vine and Wilson. The article examined the Juan de Fuca Ridge in detail and furnished a more convincing example of symmetrical anomalies than Wilson's article had offered. Moreover, its approach was quantitative. Vine and Wilson related the chart of reversals of the earth's magnetic field, calibrated in millions of years, to the zebra-skin pattern on the ocean floor, measured in kilometers, and deduced spreading speeds from it.

They identified anomalies corresponding to the Brunhes, Matuyama, Gauss, and Gilbert epochs and located their boundaries. They measured the distances between boundaries, which were found to correspond to precise dates on the chart of reversals and provided a means of calculating the speed of the conveyor belt and therefore the spreading rate of the ocean floor. They obtained figures on the order of a few centimeters per year, the same order of magnitude that Hess had estimated through general principles and that was now substantiated. The technique of measuring spreading rates by magnetic anomalies became one of the foundations of modern geology.

The Lamont group's conversion to the drift theory began when Walter Pitman realized that the map of anomalies in the South Atlantic and the one made by Lamont and the American Office of Naval Research for an area south of Iceland could be explained very well by the Vine-Matthews-Morley hypothesis. It is possible that this sudden realization was also catalyzed by the location of the researcher's office at Lamont! The office next to Pitman's belonged to Neil Opdyke. There, mounted on the wall, was the chart of reversals to which Opdyke had made such a decisive contribution. The friendship of the two men and their daily contact no doubt helped convince Pitman: the idea was in the air, and suddenly things clicked. Galvanized into action by Pitman's intuition and in spite of Doc Ewing's skepticism, the powerful Lamont machine proved very effective in exploiting the methods of magnetic chronology. After accumulating thousands of kilometers of profiles, on the conviction that magnetism would be very important someday, Jim Heirtzler and his group finally had the missing link in their possession. Perseverance had been rewarded!

If the zebra skin is the horizontal projection of the scale of magnetic reversals, it can also be used to date the ocean floor. It is therefore possible to examine the *rate* of spreading, which seems to remain constant for a section of a given ridge and to vary from one area of the earth to another, ranging from 1 to 15 centimeters per year. For example, the South Atlantic Ridge has a speed of 2 centimeters per year, the Nazca Ridge, a branch of the Pacific Ridge off the coast of South America, a speed of 12 centimeters per year. It takes 100 million years to form 2,000 kilometers of ocean bottom at a speed of 2 centimeters per year. Paleomagneticians had established the scale of reversals for only the last 4 million years, but the alternation of positive and negative anomalies extends over almost the entire ocean floor, so a chronological scale of reversals had to be *invented* for periods previous to those studied by Allan Cox. Heirtzler and his team accomplished this, using the hypoth-

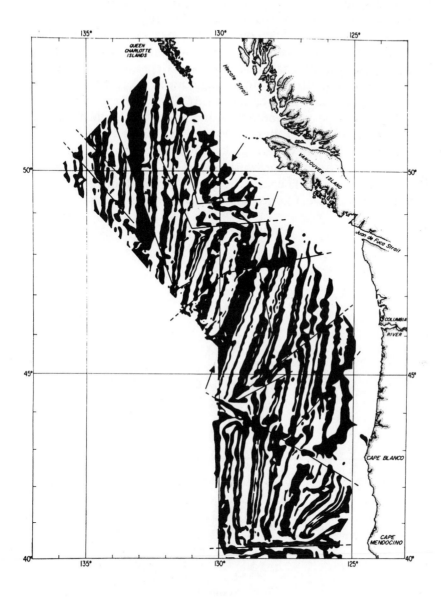

FIGURE 25   It was this map of the Juan de Fuca and Gorda spreading centers that gave Fred Vine and J. Tuzo Wilson evidence of the detailed magnetic structure of an oceanic plate. Note the numerous faults (indicated by straight lines). Without a detailed map the complexity created by these faults would hide the simple origin of marine magnetic anomaly stripes.

esis that the spreading rates observed for the last 4 million years had remained constant over time for each region.

The Lamont group first compiled the magnetic data from a large number of ocean crossings. For each crossing on a path perpendicular to the ridge the magnetic anomalies were noted, starting from the ridges, and little by little their forms were compared and correlations established among them. Thus the group identified what it called positive anomaly 1, anomaly 2, and so on. That done, they tried to date each anomaly by stratigraphic methods, wherever possible. When they were lucky enough to obtain a sample of sediment in direct contact with basalt, they were able to attribute a stratigraphic age (Eocene, Oligocene, Miocene) to it on the basis of the fossils it contained. Since absolute chronology makes it

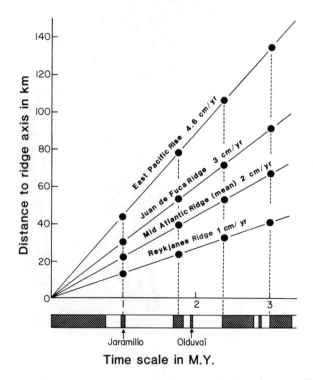

FIGURE 26 We use magnetic anomalies to calculate the spreading rates along different mid-oceanic ridges. In faster-spreading ridges there will be greater distances between anomalies. Here we plot the distance from the ridge axis of several recent magnetic reversals for four spreading centers. The points lie along straight lines, showing that these ridges have been spreading at different but constant rates for the last few million years.

possible to establish the relationship between geologic epochs and the chronological scale expressed in millions of years, the age of the sediments, and therefore of the basalts underlying them, could be estimated in this way.

By using successive approximations the researchers, who had a great variety of examples at their disposal, were able to propose a scale of reversals that went back 80 million years to the Cretaceous! (Allan Cox and Richard Doell's scale went back only 4 million years.) On the basis of this information, they attributed an age to each geographic region of ocean bottom and thus completed the geologic map. Each section of the ocean was therefore characterized according to the geologic period in which its basaltic floor was formed. The most obvious result of this mapping was the confirmation of the idea that the rate of seafloor spreading is not uniform for all the oceans. Apparently these differences in spreading rate have been maintained for almost 80 million years.

The Age of Oceanic Sediments

Now we arrive at the second consequence of the conveyor-belt hypothesis: since basalts are formed at the ridges, if the seafloor is spreading out continuously the sediments that are laid down on top of the basalt must increase in age as they move with the floor away from the ridge. To verify this hypothesis cores of sediment that had been bored down to the basalt—cores 500 to 1,000 meters in length—were systematically collected. Keeping a drill bit and shaft on even keel in rough seas at a depth of 4,000 meters is no simple matter; this was a real technical accomplishment! No doubt the possibility of improving the techniques of deep-sea drilling for oil occurred to the promoters of the program involved in this research, called JOIDES (Joint Oceanographic Institutions for Deep Earth Sampling), which united five American oceanographic institutions under the aegis of the National Science Foundation. The results fully confirmed the proposed hypotheses. The age of the sediments in contact with basalt increased regularly as their distance from the ridges increased. The rule of growth obtained by this method agreed with that deduced from the magnetic anomalies. It varied from the different oceans, and the variations corresponded to the variations in the spreading rates.

Deep-sea drilling furnished results that went far beyond the simple confirmation of seafloor spreading; it provided the foundations for the study of ocean stratigraphy. It is worth noting that the deep-sea drilling project was the first large-scale program employ-

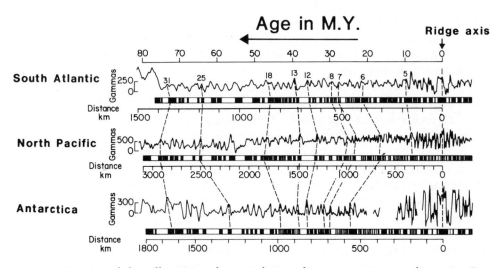

FIGURE 27 Actual data illustrating the use of anomalies to measure spreading rates. Jim Heirtzler and his colleagues correlated the magnetic anomalies of three different spreading centers for the past 80 million years. The variable distances between anomalies indicate the different rates of spread over this period.

ing an expensive technology to be put into effect in the earth sciences. No doubt it was no accident that it took place at a time when the Apollo program was reaching its zenith and people were beginning to grasp the fact that the era of cheap and abundant oil was coming to an end. Still second to physics, with its costly giant accelerators, earth science was beginning to be considered in research budgets!

The existence of the oceanic conveyor belt is well established now. For the first time, we have a framework into which all the observed phenomena of marine geophysics fit coherently. The consequences of this theory are of two types: oceanographic and geophysical on the one hand, and geodynamic on the other.

Heat and Morphology

The most obvious feature of the ocean bottom is the existence of ridges and deep trenches. According to the theory of seafloor spreading, ridges are created by the ascent of hot and therefore expanding matter along a zone in which material from the mantle is rising. As the floor spreads out on both sides, in permanent contact with seawater at a temperature of 4° Centigrade, it cools and contracts.

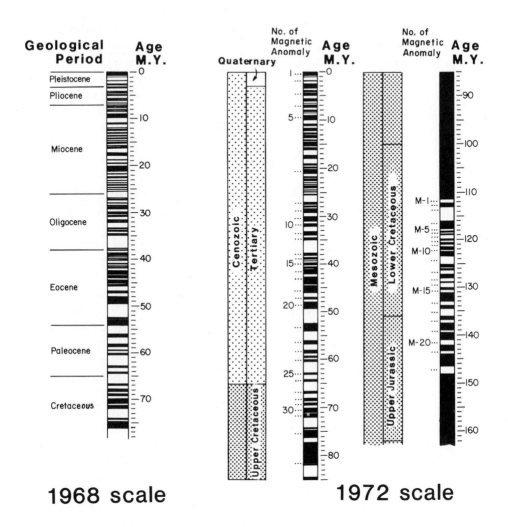

Geological Period	Age M.Y.		Age M.Y.		Age M.Y.

1968 scale

1972 scale

FIGURE 28   From seafloor magnetic anomalies and some stratigraphic data we can map a more detailed magnetic time scale than that established by paleomagnetic measurements. Reproduced here are two time scales, one compiled in 1968 and the other in 1972. Magnetic field reversals have a fairly general character throughout geologic time, but certain times are without any field reversals. These are called magnetic "quiet" zones and are found, for example, in the Mid-Cretaceous and Upper Jurassic. No "zebra stripes" are found on the ocean floor created during these time intervals.

Therefore, the altitude of the floor must decrease as it moves away from the ridges. Calculations show that the agreement between the profile expected as a result of theoretical thermal contraction and the measured bathymetric profiles is excellent. In contrast, at the abyssal trenches, some of which (such as the Kurile and Puerto Rico trenches) attain depths of 10,000 to 11,000 meters, the ocean floor plunges into the mantle, creating a sudden drop-off and a steep downward slope. The continual movement of the trenches prevents them from being filled with sediments.

The second observation that interests us here, much more sophisticated both in conception and in measurement than the first, concerns the distribution of the *internal heat flux*. The temperature increases toward the earth's center—rapidly enough that miners in deep mines have noticed it for centuries. Since heat moves from warm regions toward colder ones, heat from the earth's interior moves toward the surface.

The flux of heat is the quantity of heat that moves toward the surface from the interior per unit time and per unit area. Its absolute value is expressed in heat-flow units (HFU), or microcalories per square centimeter per second. The average flux at the earth's surface is one HFU, which is ten thousand times smaller than the flux we receive from the sun in daytime. What interests us here is not, however, the absolute value of heat flux, but its geographic variation. The flux is many times higher than the average value at the ridge crests, and it decreases slowly along their flanks. Given the way oceanic crust is formed at the ridges, this is natural. The mantle rises there, bringing to the surface very hot material that is usually found at a depth of several dozen kilometers. This material gives off a great deal of heat, resulting in a central heat-flow anomaly at the crest of the ridge. As the crust cools, it contracts and its altitude decreases. This is similar to what happens when a soufflé is removed from the oven: at first it is inflated, majestically high, and gives off abundant heat. If cold water were poured on it, it would collapse and contract, its volume would diminish, and it would cease to discharge much heat. This culinary disaster is a simplified picture of the way the altitude of the ocean bottom decreases as it moves away from the ridge and cools down.

The variation in heat flux and submarine morphology are two manifestations of the same cooling phenomenon. The pioneering work of K. E. Torrance and Don Turcotte at Cornell and of Dan McKenzie, John Sclater, and Jean Francheteau, then at Scripps, explained and made use of this phenomenon. When a ridge, such as the Mid-Atlantic, is spreading slowly, cooling takes place when the

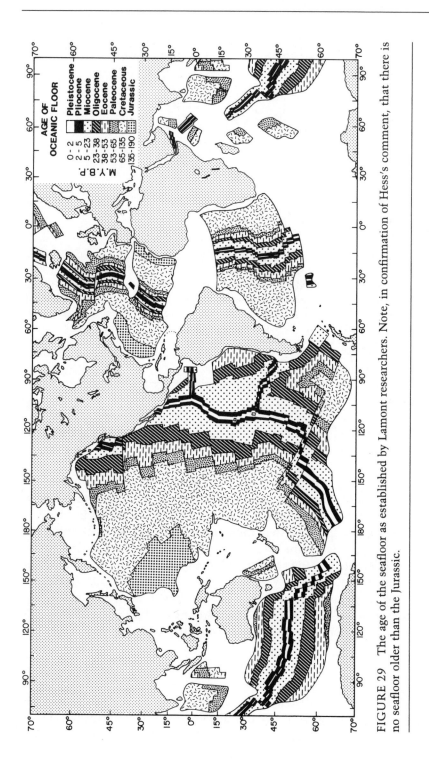

FIGURE 29 The age of the seafloor as established by Lamont researchers. Note, in confirmation of Hess's comment, that there is no seafloor older than the Jurassic.

AGE OF
OCEANIC FLOOR

		M.Y.B.P.
■	0 - 2	Pleistocene
	2 - 5	Pliocene
⣿	5 - 23	Miocene
⫝̸	23 - 38	Oligocene
☰	38 - 53	Eocene
	53 - 65	Paleocene
⣿	65 - 135	Cretaceous
⠿	135 - 190	Jurassic

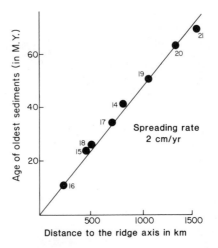

FIGURE 30 The JOIDES program of drilling the ocean floor in the Atlantic has succeeded in penetrating the sedimentary layer to the basalt basement. Using micropaleontology, researchers have determined the age of the sedimentary layer directly in contact with the basement. Plotting this age versus distance to the ridge crest, we can obtain the spreading rate. The rate plotted here is 2 centimeters per year, which is in agreement with results obtained from examination of magnetic anomalies.

hot material has not moved very far from its formation at the ridge. It reaches thermal equilibrium and sinks to a low altitude near the ridge. In other words, the oceanic crust created at the ridge descends rapidly from 1,000 meters to 4,500 meters below sea level. These ridges are characterized by high relief and steep slopes. On the other hand, when a ridge, such as the East Pacific Rise, is moving rapidly, the phenomenon works in the opposite way. The crust remains warm and inflated even after it has traveled a considerable distance from the ridge crest. These ridges are therefore less steep and very wide. Thus, the Pacific ridge's "influence zone" extends over more than 2,000 kilometers on each side of the center, whereas that of the North Atlantic ridge can be felt for only 1,000 kilometers around its central valley.

Earthquake Zones

The study of the distribution of earthquakes on the earth's surface, seismic geography, supplied a new argument to confirm the theory of seafloor spreading. The world map of earthquake activity clearly

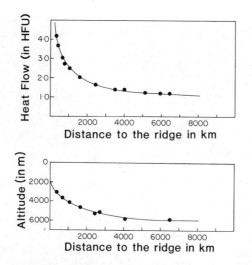

FIGURE 31 The measured values of heat flow and seafloor depth versus distance to the ridge axis correspond quite closely with theoretical calculations of a horizontally moving plate that cools as it moves away from the ridge axis.

shows the existence of two zones with high concentrations of earthquakes: the mid-oceanic ridges and the oceanic trenches at the edges of the great oceans. K. Wadati and Hugo Benioff showed that deep earthquakes were located exclusively at ocean trenches. The periphery of the Pacific is surrounded both by trenches and by a very active seismic belt. The perimeter of the Atlantic has no trenches except in two places—the Lesser Antilles island arc and the Sandwich island arc—and Atlantic earthquakes are found specifically in these regions. Both the Atlantic and the Pacific ridges show strong seismic activity.

Earthquakes are the manifestation of sudden and unrestrained ruptures, the earth's most spectacular example of the liberation of mechanical energy. The creation and destruction of the ocean floor produce mechanical strains that take the form of earthquakes. The theory of continental drift thus provides a unified framework for all the oceanographic observations: magnetism, morphology, seismicity, heat flux, and the distribution of sediments. So it is not surprising that the first scientists to use this theory were marine geologists and geophysicists.

The Return of Continental Drift

The most spectacular consequence of the spreading of the ocean floor is no doubt continental drift, the starting point of Wegener's

John Sclater

theory. The continuous creation of the Atlantic floor at its central ridge is conceivable only if the continental borders of the ocean have moved apart from each other. From this observation came the idea of drift, which turned out to be very similar to the one that had been postulated by Wegener. Now, thanks to magnetic dating of the ocean floor, a precise chronology of the history of drift can be established by superimposing the successive contours of magnetic anomalies of the same age on each side of the ridge. For example, anomalies 2 are superimposed one upon the other, so are anomalies 3, and so on. Every "piece" of crust is automatically displaced the same amount, including the continents, which are only following in the path of the immediately adjacent shore. The chronologic chart of reversals becomes a remarkably effective tool for reconstructing the evolution of oceans, because the magnetic anomalies are clearly symmetrical to the ridge and can therefore be recog-

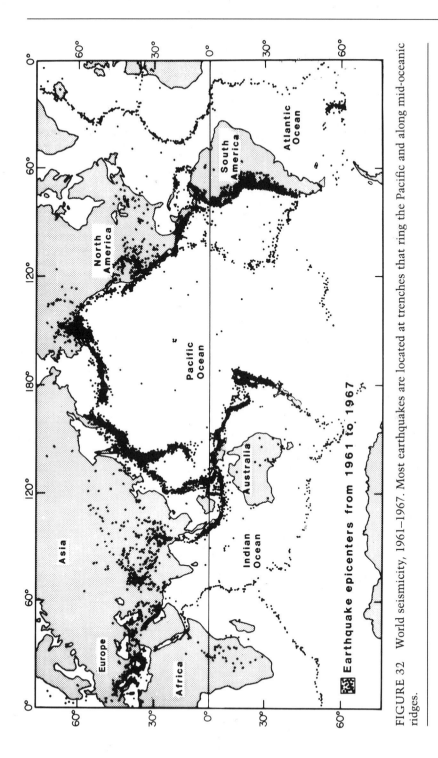

FIGURE 32   World seismicity, 1961–1967. Most earthquakes are located at trenches that ring the Pacific and along mid-oceanic ridges.

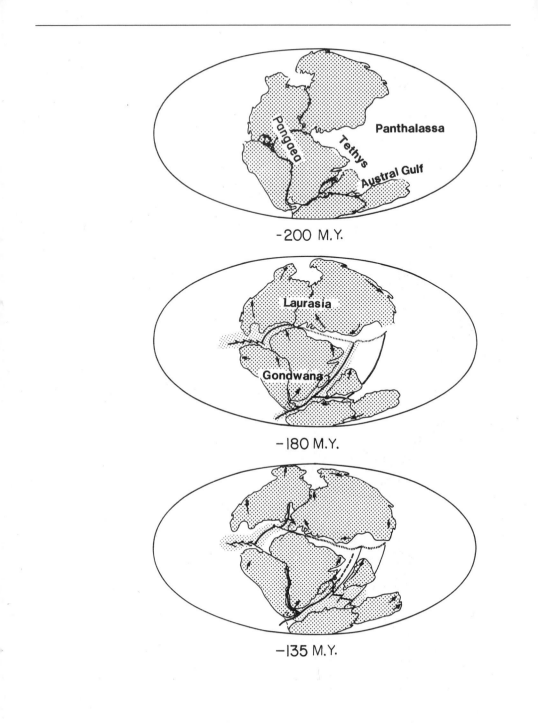

-200 M.Y.

-180 M.Y.

-135 M.Y.

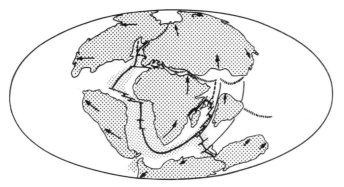

− 65 M.Y.

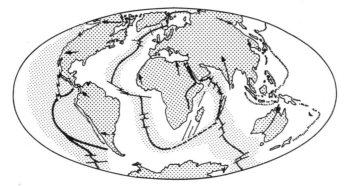

0 M.Y.

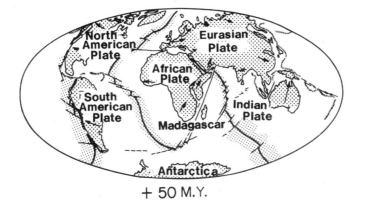

+ 50 M.Y.

FIGURE 33 Dietz and Holden based this reconstruction of the breakup of Pangaea on records of seafloor magnetic anomalies. Note the similarities between this map and those made by Wegener and by paleomagneticians. Note, also, that Dietz and Holden had enough confidence in their method to predict what the earth would look like 50 million years in the future.

nized, identified, and dated. For example, maps of the anomalies show that the separation of North America and Africa began at the end of the Triassic (200 M.Y.B.P.) and that the North Atlantic drift took place much later, beginning in the mid-Eocene (60 M.Y.B.P.). The history of the oceans was reconstructed in this way and a scale established by Jim Heirtzler and his colleagues at Lamont, who were able to map Pangaea and its breakup on a much more rigorous base than was possible in Wegener's time. The picture that emerges from this reconstruction is rather different from Wegener's, not in its main outlines but in some specific events. To see this one could profitably compare the maps prepared by Robert Dietz and John C. Holden to Wegener's (see Figure 4).

A Unified Earth and a Relativistic Geology

The rebirth of continental drift, the magnetic mapping of the ocean floor, the determination of seafloor spreading rates, and the results of seismic geography all suggest that the earth is a complex machine with complex dynamics. Its surface is being constantly changed and renewed, and its workings can be understood only from a global perspective. It is not possible to restrict research to a particular area without risk of misunderstanding its evolution; scientists must, therefore, interest themselves in the earth as a whole. If Africa and Europe move closer together, repercussions will be seen in Australia or Japan, because such movement will change the spreading rates or the orientation of the adjacent ridges. If a subduction zone appears off the coast of Peru, all movements in the Pacific will be affected by it. The law of simple and immediate causes and effects must be replaced by a law of complex systems in which causes are also effects, in which everything is interrelated, and in which everything plays a part in the behavior of the earth. Loops and feedback systems preclude a hierarchical or unidimensional description, so it is necessary to examine the whole. Regional geology, which was as overpolished as a painting by Ingres, must henceforth be integrated into a larger whole; perhaps not every parameter must be taken into account, but at least those principal ones that define the functioning of the terrestrial system.

Mobility introduces necessary complications into this global view. Let us look at the behavior of Africa during the last 200 million years. The continent has been "wedged" between two spreading centers, the Indian and Atlantic oceanic ridges. One might expect that it would be subject to intense compression, but eastern Africa contains a great rift valley marked by huge volcanoes (such as Kilimanjaro) and imposing lakes that give evidence of strong extension. Some opponents of drift theory even used this example as proof of its falsehood. In fact, everything can be explained if we take into account that the distance between the Indian and Atlantic ridges is not fixed but is increasing over time.

The ridges are clearly mobile, not fixed. Like a ballet master

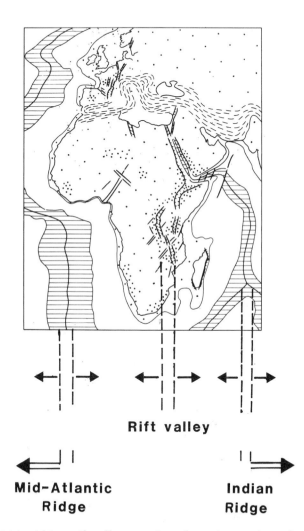

Rift valley

**Mid–Atlantic
Ridge**

**Indian
Ridge**

FIGURE 34   African rift valleys are sites of ongoing continental extension. Even though Africa is surrounded by mid-oceanic ridges, extension and not convergence is occurring there.

who, having choreographed a step, is swept into the dance, they are pulled along in compensation for movements in underlying layers that are themselves driven by the phenomena of spreading. All is in movement on the surface, and one must become accustomed to looking at these movements relatively, not in the absolute sense. America is not moving while Africa stands still; nor is it the other way around. Both continents are moving with respect to each other. In the 1950s paleomagneticians described the waltz of the continents and, to avoid getting dizzy, chose the magnetic pole as a fixed reference point. Even so, they made the pole migrate back and forth rapidly, obliging anyone looking for an absolute landmark to work twice as hard. The search for an absolute natural landmark was pursued by drift specialists, because mechanics, which is full of mobile landmarks, is very difficult to master. But here, as in fluid mechanics, you have to work with what you've got!

Thus the earth's surface is like a bubbling liquid, continuously in motion, pulling the unsinkable continental rafts along. Unfortunately, the motion of this liquid cannot be described in relation to a fixed landmark with a clear geologic significance. Only the surface fluctuations are known. A person seated on a train and looking at another train can determine which train is moving only by looking at a fixed point: a tree, building, or post. In our own case, being pulled along on our continental rafts, we have no fixed point, so we must be content to describe the movements of continents and oceans in a *relative* way.

Geologists continued to search for the missing reference point. Only with the development of the hotspot theory could absolute reference points be defined geologically. But more on that later.

Arthur Holmes

The perspectives offered by the theories of seafloor spreading and continental drift are so illuminating that one wonders why they did not resurface sooner. One marvelous gem of thought did exist, however, and it was in its own way almost as perfect as Harry Hess's article. This was Arthur Holmes's beautiful treatise of 1945, *Principles of Physical Geology*. The Scottish pioneer developed the idea that oceanic ridges are traces of currents ascending through the mantle, elements in a vast convection system. Like Hess, he postulated the spreading of the ocean floor and lateral continental drift. Holmes's book was read by all students in England and recommended by all professors, but apparently none of them attached much importance to the drift theory. Englishmen like

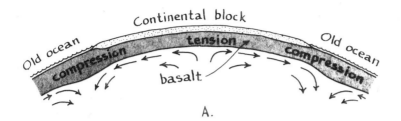

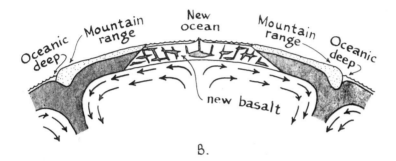

FIGURE 35 In 1945 Arthur Holmes proposed a model that in essence was the debut of seafloor spreading. All modern concepts of spreading are evident in this figure.

Fred Vine and Dan McKenzie would have to go to America to find their inspiration, although since Holmes and Runcorn the idea of continental drift had not left the United Kingdom! In this regard Holmes's role was more important than one would think; thanks to him, Great Britain was well prepared for the drift idea and supplied a large number of its innovative thinkers. But, during this same period, British earth scientists were made very cautious by the inflexibly dogmatic and negative attitude of Sir Harold Jeffreys! Holmes was a geologist; Jeffreys, a geophysicist: how could young geophysicists in Great Britain let down their own side?

No doubt the major reasons Holmes's ideas were not considered more seriously was that he offered no new arguments to buttress Wegener's theory. This theory had been "proved false": why go back to it? If we keep this in mind, it seems clear that the "click" of understanding was inspired not by Hess, but by Vine and Matthews, who developed a powerful argument supported by tangible facts. Fred Vine's achievement was not so much to have developed a new interpretation as to have been the first to *convince* others. J. Tuzo Wilson's support would be decisive. In a few years, almost alone, they would persuade everybody.

Which is more important in science: to discover or to convince? Ideas and discoveries evolve progressively and generally without major lacunae between them (many people feel that the difference between Morley-Vine-Matthews and Matuyama is rather minor). The ideas of Wegener, Du Toit, Argand, B. Choubert, Holmes, Hess, and Dietz form a perfectly continuous progression. On the other hand, the "clicks" in the collective understanding proceed by jumps or abrupt changes of direction. Sometimes these sudden leaps lead to errors, and an idea quickly goes out of vogue; sometimes they lead to the formation of a new discipline. This phenomenon takes place more abruptly in America than in Europe, but its results are similar. Like a percolation process with a threshold, ideas remain isolated and unconnected for a long time and then they join together, form a continuous network, and can be extracted as a homogeneous entity. So it was with the theory of seafloor spreading. Marine geophysicists were hostile to it; they all thought they could prove the contrary with their own observations—even the researchers at Scripps and Lamont. In a matter of months, however, the members of the oceanographic community swung in the opposite direction and became believers. They offered new arguments in favor of drift with a fervor that was as passionate as it was newfound, almost to the point of dogmatism. Like a school of fish suddenly changing direction in precise synchrony, oceanographers finally adopted seafloor spreading!

PLATE TECTONICS

4

Meetings, Publications, and Priority

Traditionally the annual meetings of the American Geophysical Union have been held in Washington in the spring. (Recently they have migrated to other cities on the East Coast.) They are the annual "fair" of the earth sciences, for which six thousand people assemble in an enormous hotel for a week. Earth scientists present the results of their latest research in about a hundred sessions—say, a dozen each day. All the "stars" of American science are there, as well as students just finishing their Ph.D.'s, established researchers, and specialists from Europe, Australia, and Japan. Discussions begun in the meetings continue in the halls and around glasses of beer, from eight o'clock in the morning until well past midnight. Everyone speaks of their most recent results and also of their plans. The young researchers who have just finished their Ph.D.'s show their "wares" in the hope of making a name for themselves and attracting the attention of—and job offers from—university department heads or directors of research for private companies. Department heads assess the quality of their own groups and look for those "rare birds" likely to strengthen them. Various commissions and committees also choose this occasion to meet and distribute funds or define programs.

In the Middle Ages tournaments that were first organized simply as competitions among knights gradually became meeting places and then centers of commerce and information. A.G.U. meetings are a modern version of the tournament—both joust and market-place of ideas for the earth sciences. The meetings play an important role in evaluating research. News of a discovery made in Massachusetts spreads immediately throughout the country, from the big California universities to the small and obscure ones in the wilds of Nebraska. It will immediately be taught to students and thus integrated into the educational system. Sometimes the year's hit discovery will turn out to be wrong. It will be corrected the following year, for it will not have survived testing by an enthusiastic but sharp and critical scientific community. Usually it will stimulate new research and help to maintain the high level of American science.

Like money, discoveries tend to depreciate in value over time, so it is important for a new idea to become known rapidly. Direct contact among specialists fosters the rapid and efficient diffusion of scientific discoveries. Experimental tricks and techniques are exchanged, failures one has not dared to publish are confided, and friendships that may lead to future collaborations are begun. In the United States this policy of direct contact is particularly favored by the system and works to the benefit of all, not just the department heads and established researchers. Graduate students in American universities are initiated early into this science of direct exchange, of public contact and personalized presentations. They learn to prepare attractive colored slides, to give intriguing titles to papers, and to make oral presentations that are short and clear. A constant flow of diverse and unorganized information excites the crowd of avid seekers of new ideas. No longer are scientific discoveries introduced in the pages of scholarly journals: one hears them first in oral presentations or, if they are really new, in rumors (whose importance is becoming a disturbing factor in modern science). By definition a rumor is unverified, always distorted, and sometimes false, but on the basis of rumor alone some scientists give up important research, thinking their work has been outmoded. Others go off on scientific wild-goose chases. Rumor fuels the sciences of "guesstimation" and of the sensational. As in the domain of technology, everything moves very fast. Publications are still the official records and references, but publication delays get longer every year: by the time research reports appear in print they are often a year or two out of phase with the current status of a project.

A collection of summaries of the oral presentations is usually provided at scientific meetings, but these *abstracts* are frequently incomplete, and their contents do not always correspond to what the author actually said. Therefore, the majority of the scientific community does not consider an abstract a reliable record, and the tangle of abstracts, oral presentations, and articles often creates controversies. An example of such a scientific imbroglio is the birth of the theory of plate tectonics.

At the A.G.U. meeting of April 19, 1967, W. Jason Morgan, at that time a young professor at Princeton, presented a talk entitled "Convection in a Viscous Medium and the Formation of Oceanic

Trenches" during a special symposium on island arcs, oceanic ridges, and seafloor spreading. It was noon. The room, which had been nearly full until then, was three-quarters empty, as often happens right near lunchtime.

Morgan's talk had only a distant relationship to the abstract printed and distributed before the start of the congress. He stated that the earth is divided into rigid plates and that the lateral movements demonstrated by Vine and Matthews can be described by the laws of spherical geometry. At the end of the nineteenth century this branch of mathematics was studied intensively by mathematicians attempting to deduce the laws governing the displacement of a point or a set of points constrained to remain on the surface of a sphere. For example, imagine that we cut a piece of skin from the surface of an orange with a sharp knife. Suppose that we move this patch of orange skin from its original place and slide it along the surface of the orange (see Figure 36). In the terms of this simplified example, the relative movement of the mobile section is a rotation about an axis passing through the center of the orange while the rest of the orange skin remains in place. Because the Swiss mathematician Leonhard Euler was the author of this theorem, the axis is called the Eulerian axis of rotation of the detached portion of the sphere A in relation to the fixed portion B.

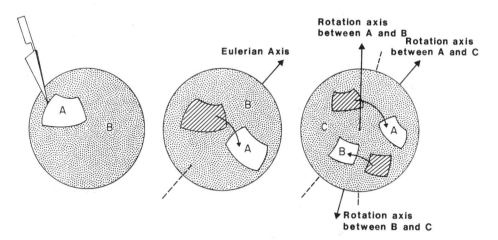

FIGURE 36   Imagine that we take an orange, cut out sections of the peel, and move the sections around. This exercise shows the relative movements of spherical shells, explained by Euler's theorem. The movement of section A over the surface of the orange is described in terms of a single rotation about a pole called the Eulerian axis. In general we measure the displacement of any part of a sphere in terms of a finite rotation about some Eulerian axis.

The complexity of the game is increased if not one but two windows are cut in the orange skin. Then we have two movable sections, A and B, and a fixed portion, C, still attached to the orange. If we slide sections A and B over the surface of the orange, we describe their relative movements—A in relation to B, B in relation to C, and A in relation to C—as rotations about an axis, but the axis is not unique; for each rotation there is a distinct Eulerian axis related to a particular relative movement.

Sliding a cut-out section of an orange skin over the fixed portion has two effects: at one boundary the inside of the orange is uncovered, and at the other boundary one layer of skin sits on top of another. Where the inside of the orange is uncovered a "new" surface has appeared; on the other edge, where the two layers of skin overlap, a part of the old surface has disappeared. Adopting geological terminology to describe the orange, we could say that the region in which the layers of skin overlap, where one piece sinks under the other, is the equivalent of an oceanic trench or subduction zone, and that at the point of contact there is a surface deformation equal to the thickness of the skin. The area in which the two pieces of skin are moving away from each other is the equivalent of an oceanic ridge zone.

Now suppose that the entire skin is cut in pieces that are then made to slide over the surface of the orange. That is almost a model for the earth's surface, except that on the earth new surfaces are *created* symmetrically on both sides of the ridges.

Using the orange analogy (including the exception) to describe the earth, Morgan asserted that the globe is composed of rigid portions of a sphere called plates. New areas of plate are formed continuously at the ridges. Once formed, the plates move rigidly, without changing their shape, across the surface of the earth until they disappear into the mantle at the oceanic trenches. The ridges are zones of creation (accretion) and the trenches are zones of destruction (subduction) of the surface. Transform faults form circular arcs, centered on the axis of rotation, that describe the movement of the two plates that they separate. The positions of these discontinuities indicate the orientation of the movement of the rigid plates. In addition, the movement of the ocean floor is measured by the axis of rotation of the movement between two plates and the speed with which the angle that describes the rotation varies over time (called the angular velocity).

Morgan used the North Atlantic as an example of how the pole of rotation between the American plate and the African plate can be determined and in so doing presented the theory of plate

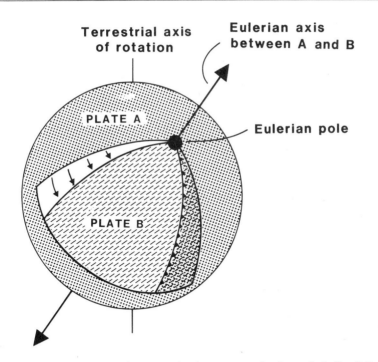

Terrestrial axis of rotation

Eulerian axis between A and B

Eulerian pole

PLATE A

PLATE B

FIGURE 37   Now consider not only the motion of spherical shells sliding over the surface of the orange but also the creation and destruction of *new* surface. In this example a triangular shell moves about an axis passing through the center of the sphere and one of the corners of the triangle. We call this the Eulerian pole of rotation between plates A and B. With this motion we expose at the left a small section of naked orange. This region where new surface is created is analogous to a ridge axis. In contrast, to the right plate B disappears beneath a part of plate A. When one plate goes beneath another we say it is "subducted." The area created at the ridge is equal to the area subducted because the area of the orange remains constant. The three sides of triangle B correspond to a boundary of surface creation (in this figure, the left side of the triangle), a boundary of surface destruction (the right side), and a boundary of surface conservation (the bottom side), which is called a transform fault.

tectonics to the public for the first time. Thus W. Jason Morgan was the father of this theory. When the written record is consulted, however, the attribution of paternity becomes complicated.

Six months after the A.G.U. meeting, in October 1967, Dan McKenzie and Robert Parker of England published an article in *Nature* proposing a theory very similar to Morgan's, but using arguments based on seismology. With the Pacific Ocean as their example, they developed the theory of "tectonics on the sphere."

Morgan's article corresponding to his oral presentation did not

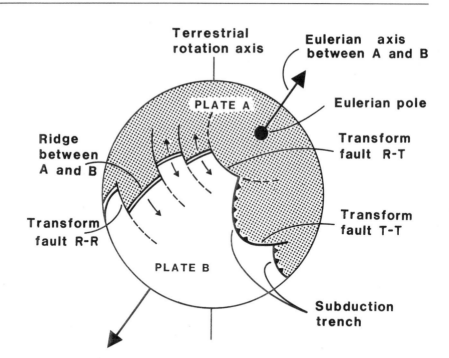

FIGURE 38   A less simplistic model of an earth consisting of two plates, A and B. Spreading occurs symmetrically on the ridge, and both the ridge and trench are segmented and offset by multiple transform faults. Transform faults are diverse in nature, but all share the fundamental characteristic of forming small circles about the Eulerian pole of relative motion between plates A and B. (Note that the Eulerian pole that describes the movement of A and B has nothing to do with the earth's axis of rotation through the North Pole.) As shown here, transform faults may border two ridge segments (R-R), two trench segments (T-T), or a ridge and a trench (R-T).

appear until February 1968. The article had been submitted in the spring of 1967, but it remained in the hands of a reviewer for half a year, had to be revised, and thus was accepted for publication only after a long delay. This article, which had been submitted before all the others on the movement of plates, was very complete. It did not simply state principles, like McKenzie and Parker's article. It presented numerous calculations of trajectories of rotation using abundant data on transform faults. Although difficult to read, it was a powerful, profound, and prophetic contribution, in spite of its late appearance. Because Morgan's article was delayed a full year, a chronology based on publication dates would give priority to McKenzie and Parker.

It should be possible to prove that Morgan was first by referring to the printed abstract that accompanied his oral presentation at the A.G.U. meeting. Unfortunately, the abstract of Morgan's talk is of no help, because it does not correspond to his oral presentation! The subject of the abstract Morgan sent to the A.G.U. in early January 1967 was trenches. At the end of January, however, *Science* published an article by Bill Menard, an oceanographer at Scripps, concerning fracture zones in the Pacific Ocean. Menard showed that fracture zones defined small circles, not large meridional circles, as had previously been thought. For Morgan, this assertion triggered an idea; later he said, "In five minutes I understood it all." Even though his abstract had already been sent out, he decided to draft an article, submit it for publication, and present it at the A.G.U. meeting. He had no way of knowing that it would take so long for the article to be published! He also couldn't know that his oral presentation would eventually be considered so seminal.

But we have not yet come to the end of the story. Dan McKenzie was present at the A.G.U. meetings in 1967. He attended the session during which Morgan was supposed to give his talk, but he left the room just before it began. So he didn't get wind of Morgan's idea, and his own approach must be considered an independent

Bill Menard

one. This interpretation of the facts, given by McKenzie himself, seemed suspicious to many people who thought he must have heard talk of Morgan's discovery either at the A.G.U. or at Scripps, where he was working at the time. Morgan himself thought that McKenzie's version of the story was probably true, however odd it might seem. (I thought so, too.) "Nobody," Morgan said, "attached any importance to the idea of plates, so why would it have been talked about? Besides, my talk was scheduled at noon. Dan, like the majority of the audience, had left for lunch by then!"

Thus the central idea of plate tectonics, which was to apply the mathematical theorems describing the motion of sections of a sphere to geologic formations on the earth's surface, sprouted independently in (at least) two minds within a period of a few months. This example, which is not unique, shows that when an idea is ripe, its formulation is practically inevitable and often depends on secondary factors.

Are these accidents of fate important? Not if one looks at the evolution of science from the remote distance of the star Sirius without worrying about the protagonists or the vicissitudes of their lives. But they are important if one wants to understand how scientific research progresses from day to day, if one is interested in the sometimes winding paths by which ideas are propagated. And, again, they are important in understanding the *sociology* of science. After all, prizes, laurels, promotions, and especially recognition of paternity play a fundamental role in research! One might wish it otherwise, preferring to see only an enlightening progression of ideas in the development of scientific theories and to leave in obscurity those who gave birth to those ideas and were responsible for their triumph. The motivations of scientists, without whom no idea would be born or developed, are no different from those of other creative people. They all want to be recognized and to have it known that their contribution to scientific progress was significant. And they all live in a perpetual state of self-doubt. The famous French biologist Jacques Monod, who had just received the Nobel Prize, once recalled the agony of the researcher: "In research self-satisfaction is death! Doubt motivates progress, but it is painful to endure."[*] In the context of the real world of research, in which the notion of discovery assumes a very high value, it should be remembered that paternity is not always easy to establish.

Jason Morgan's oral presentation did not pass completely unnoticed, however. Xavier Le Pichon of France, who was present and

[*] In an interview in *Nouvel Observateur* in 1965.

listening carefully, immediately saw how he could make use of Morgan's theory. He had in his possession a set of data on the magnetic anomalies of the ocean floor and their chronological occurrence. He had just finished working up this data with Jim Heirtzler's group at Lamont and knew the exact locations of the transform faults. Consequently, he could reconstruct the relative movements of the continents, using the concepts of spherical geometry, rigid plates, and axes of rotation as they had just been developed by Morgan.

Le Pichon made rapid progress and was luckier than Morgan. His article was accepted by the *Journal of Geophysical Research* and published in March 1968. In it he divided the earth into seven plates, defined the borders of each one, and calculated the poles of rotation for the relative motions. He designed the first plate-tectonic diagram of continental drift and calculated the relative movements of the continents over the past 200 million years.

Le Pichon's approach was less original and theoretical than Morgan's or McKenzie and Parker's, but it was more systematic and more global. The two approaches were not independent, however; as Le Pichon himself stressed in his introduction, his contribution was but an application of Morgan's theory. Only the chance proximity of the publication dates suggests that these two historically sequential steps should be grouped together.

Learning from Earthquakes

Earthquakes and volcanoes unleash the internal energy of the earth in a most sensational manner. The clearest statement of the world model of seismological synthesis was presented by Bryan Isacks, Jack Oliver, and Lynn Sykes of Lamont in an article published in 1968 by the *Journal of Geophysical Research*. At that time Oliver and his students, Isacks and Sykes, were studying the seismicity of areas in the western Pacific, such as the Tonga and Kermadec islands, where subduction takes place. Like everyone else at Lamont, they were also interested in the oceanic ridges and their seismicity. They were familiar with the work scientists at Lamont had done on magnetism, and it aroused in them an interest in drift theory. As seismologists, they were trying to fit the new theories into the framework of their own field. Let us outline their reasoning, starting with a simple observation that I have already mentioned but that I will now elaborate upon: the geographic distribution of earthquake sites (called seismic geography).

Earthquakes are not distributed uniformly or by chance on the

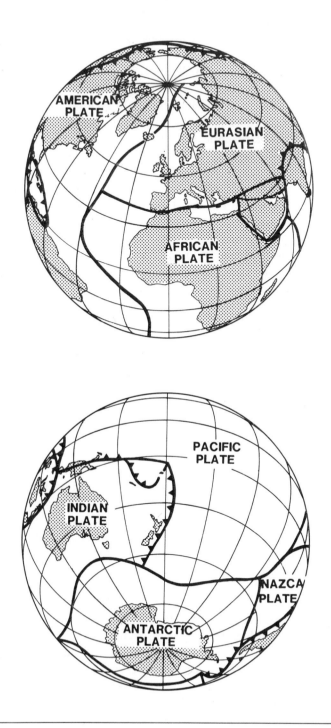

PLATE TECTONICS

FIGURE 39   Xavier Le Pichon's 7-plate description of the earth, which is extremely similar to Morgan's 9-plate version. Later work has refined this model (for example, separate North and South American plates, with slight convergence between them, have been introduced), but the basic outlines are still considered essentially correct.

The African plate is bounded to the west by the Mid-Atlantic Ridge and to the south and east by the Southwest and Central Indian ridges. The northern boundary is more complex: a ridge in the Red Sea; a series of transform faults branching from the Red Sea up through the Dead Sea and the eastern border of the Mediterranean; and a broad region of convergence and strike-slip motion through the northern Mediterranean, Gibraltar, and to the Azores, where it intersects the Mid-Atlantic Ridge.

The Eurasian plate is bordered on the west by the Mid-Atlantic Ridge, on the east by a series of subduction zones that stretch from the Kuriles to Indonesia. This convergence zone then extends westward through the Himalaya and Iran to link up with the northern Mediterranean convergence zone that abuts the African plate.

The Indian plate is bounded to the west by the Central Indian Ridge and to the south by the Indian-Antarctic Ridge. The northern and eastern oceanic boundaries of the Indian plate are trenches stretching from New Zealand through New Guinea to Indonesia and Malaysia. These link up on land with the Himalayan-Iranian convergence zone. Note that India and Australia move together on the same plate and also that the intertwined mutual subduction of the Pacific and Indian plates has led to a somewhat complicated boundary geometry.

The Pacific plate underlies most of the Pacific Ocean. It is bounded on the west by a string of subduction zones and on the east by the East Pacific Rise, which is interrupted between the Gulf of California and the Juan de Fuca Ridge by the San Andreas transform fault.

The Antarctic plate is totally surrounded by ridges.

The American plate is bounded on the east by the Mid-Atlantic Ridge. The western boundary, at the western edge of the American continent, is geologically quite complex, being a zone of trenches stretching from Alaska to Chile offset by the San Andreas transform fault and a spreading center in the Gulf of California.

Between South America and the East Pacific Rise Le Pichon placed *the Nazca plate.*

Since this model was proposed several refinements have been made, generally affecting small regions of the earth. Besides separate South and North American plates and a division of the Nazca plate into a southern Nazca plate and a northern Cocos plate (which extends from the East Pacific rise to subduct beneath Mexico and Central America), a separate Arabian plate (stretching from the Red Sea to Iran) and a separate Caribbean plate (bounded by trenches on east and west) have been found to fit geophysical observations and reconstructions of past plate motions. The earth is now thought to be divided into 13 plates.

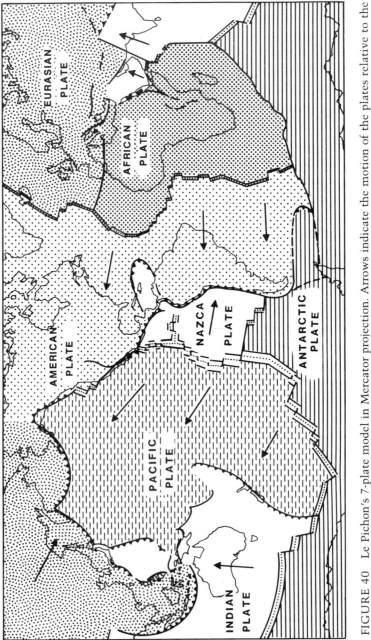

FIGURE 40 Le Pichon's 7-plate model in Mercator projection. Arrows indicate the motion of the plates relative to the African plate, which is assumed fixed by convention. Double lines separated by a dotted line indicate ridges, serrated lines show subduction zones.

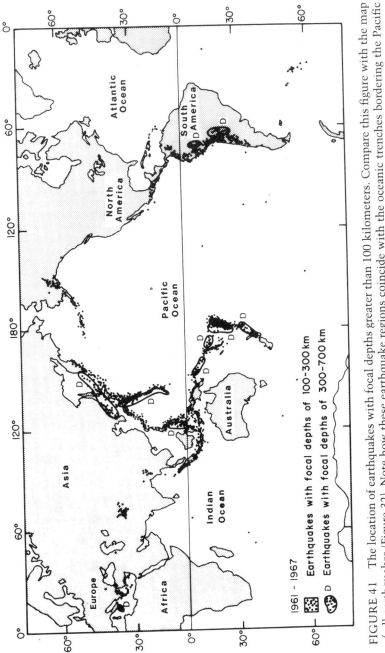

FIGURE 41 The location of earthquakes with focal depths greater than 100 kilometers. Compare this figure with the map of all earthquakes (Figure 32). Note how these earthquake regions coincide with the oceanic trenches bordering the Pacific and the Caribbean.

1961 - 1967   ▨ Earthquakes with focal depths of 100-300 km

⬭ D Earthquakes with focal depths of 300-700 km

earth's surface; they are located in well-defined zones—along the oceanic ridges, on transform faults, or in subduction zones. In the first two instances their foci are superficial, less than 100 kilometers in depth; only in subduction zones do the foci reach a depth of 100–700 kilometers. There are no earthquakes deeper than 700 kilometers, and there are no deep earthquakes outside of subduction zones. Examples are the trenches along the western edge of the Pacific, from the Kurile Islands through Japan to the Marianas, and along the eastern edge, from Mexico through Peru to Tierra del Fuego. These zones are infamous for their earthquakes and the disasters, often involving thousands of casualties, associated with them.

Earthquakes, remember, are manifestations of a more fundamental phenomenon: the sudden rupture of rigid, solid material. In general, earthquakes are produced only in the top 100 kilometers of the earth's crust, because the rigid section on which forces are exerted is about 100 kilometers thick. This rigid section is called the lithosphere. In Chapter 2 I explained that the earth consists of concentric spheres: crust, mantle, and core. How does the lithosphere fit into this picture?

The lithosphere consists of the crust (oceanic and continental) and the upper part of the mantle (the part located above the asthenosphere). Whereas the boundaries between the crust and the mantle and between the mantle and the core are marked by a change of chemical composition and each unit—crust, mantle, and core—can be characterized by its chemical properties and its mineralogical composition, the concept of a lithosphere is a purely mechanical one. The lithosphere itself is chemically heterogeneous; it is a mechanical entity: the external layer of the globe that reacts to mechanical force in a rigid manner.

No doubt earthquakes are located at plate boundaries because displays of energy are located there. Only the plate boundaries are geologically active, because they are the only places where plates interact with one another. If the oceanic ridges make the plates and therefore "create earth surface," there must be some mechanism that causes surface to disappear, since the earth's surface area is assumed to be constant. (After World War II S. Warren Carey tried to revive the idea of continental drift by suggesting that the earth is expanding continually. No objective observation suggests such a mechanism, however, and the idea met with little success.)

Seismologic study of the great oceanic trenches showed that earthquakes are located in narrow bands in the mantle about 50 kilometers thick. Doesn't that suggest the burial of a plate about

FIGURE 42   The Isacks-Oliver-Sykes model of plate configurations. Plates are created from the asthenosphere at mid-oceanic ridges, cool, and thicken to 70–150 kilometers to form the lithosphere as they slide over the asthenosphere and move away from the ridges; they reenter the asthenosphere at subduction zones.

50 kilometers thick? Subduction zones are clearly the areas into which plates disappear. The logic of seafloor spreading indicated the existence of such zones, and seismic observations located them in the areas of oceanic trenches.

The distribution of earthquakes in ridge zones has a very characteristic form, also. In ridge zones one finds swarms of rather low-intensity quakes. The transform faults that separate ridge sections are the seat of much more powerful quakes. This clearly indicates that the major mechanical force is concentrated at those breaking points called transform faults. The little ridge-ridge faults are not the only seismically active ones; the great trench-ridge and trench-trench faults are just as active. The great Mediterranean transform faults on which the earthquakes of Yugoslavia, Greece, and Turkey are found and the Chaman Fault connecting the Himalayan subduction zone to the Indian Carlsberg Ridge are well known. But the most famous transform fault is no doubt the San Andreas Fault connecting the Gulf of California and the Juan de Fuca Ridge. The great earthquakes that give California its reputation as a dangerous earthquake zone (such as the one that damaged San Francisco in 1906) are located on it.

Earthquakes are ruptures in the earth's crust, and they also emit acoustic waves. From the study of seismic waves recorded by the network of seismographs located around the world the physical characteristics of the materials crossed are determined. The physical parameter considered by Isacks, Oliver, and Sykes is the quality-of-material factor, or Q factor. When the Q factor is high, the medium is rigid and transmits seismic waves well; when it is low, the medium is more viscous and transmits waves badly. Detailed analysis of seismic records shows that over almost the whole earth the first 70 to 100 kilometers of crust has a high Q factor and the area underlying this layer, more than 200 kilometers thick, has a low Q. This observation agrees with the theory that the first 70–100 kilometers constitute the rigid plates and that the substratum is the viscous medium on which they glide. Seismologists call the upper layer with a high Q the lithosphere and the lower layer with a low Q the asthenosphere (*lithos* is Greek for "rock," *asthenos* is Greek for "weak").

Let us look at the Q factor at the plate boundaries. In the area of transform faults the factor remains the same on both sides of the fault. The same cannot be said of the area around ridges and trenches. The plastic, deformable mantle on which the rigid plates move "rises" toward the surface at the ridges (as we would expect, because the rigid lithosphere is created in these zones). At the trenches, as would be expected, a high-Q zone is submerged into the low-Q asthenosphere. We can say that the subduction of lithospheric plates has been demonstrated.

The third argument developed by Isacks, Oliver, and Sykes, which had also been used by McKenzie and Parker as a base for their demonstration, concerns the mechanisms at work at the foci of earthquakes. Because seismic activity involves rupture along a fault, one can determine the type of fault created by the earthquake by using the information recorded at observatories around the world. If compression produces the earthquake and rock is pushed upward, the fault is reverse; if the fault is produced by stretching and rock falls downward, the fault is normal; if the quake takes place at such a sharp angle that the displacement produced is strictly horizontal, a strike-slip fault occurs (see the Glossary). The mechanisms at work at the focus, in the various seismic zones, show charcteristics that are completely consistent with the plate-tectonic model. Little quakes located on a ridge are of the stretching type. Nothing could be more normal, given that surface is being created there and the plate edges are moving away from each other. On transform faults (or strike-slip faults, where displacement is horizontal) the mechanism is of the shearing type. Moreover, the direction of shear corresponds to the direction of relative motion determined by magnetic anomalies. Thus the interpretation that plate tectonics provides, based on earthquake data, is identical to that furnished by the magnetic anomalies that record long-term movement whose speed is measured in centimeters per year. In subduction zones the focus mechanisms are more complex and more varied. I will return to them later; at the moment I need only add that they do not appear to contradict the model of a plate that bends and plunges into the mantle.

Isacks, Oliver, and Sykes arranged practically all the available seismological data into a coherent, homogeneous synthesis. Their article, to which I devoted such a long explanation, worked like a detonator on the world community of geophysicists. It was certainly not the first article on plate tectonics, but it converted the largest number of adherents. Morgan's presentation at the A.G.U. meeting in 1967 passed unnoticed, but a few months after the article by Isacks, Oliver, and Sykes "plate tectonics" was on everyone's lips. Why this sudden interest? In my opinion the reason can be found in the mentality of the geophysicists at that time.

For generations teachers of geophysics all over the world had been explaining the world map of earthquakes, their geographic locations and their distribution in depth. For generations seismic cartography had been presented in courses on seismology as a separate set of data, which could not be omitted because it consisted of a great many irrefutable observations. However, the object of these courses was to study the propagation of seismic

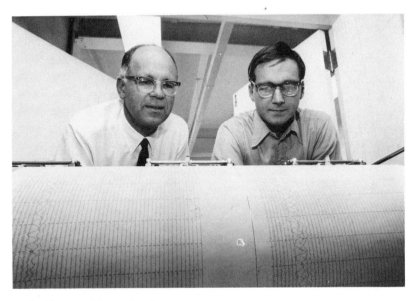

Jack Oliver and Bryan Isacks

waves. These theories are well known and are based on a solid mathematical foundation. Seismic geography, on the other hand, lacked a quantitative foundation; it was a descriptive field at the edge of science teaching. So it is easy to understand why this synthesis by the Lamont seismologists, which finally introduced order into seismology courses and returned a sense of unity to the discipline, evoked such a positive reaction. The warmth of its reception was enhanced by the fact that ocean-floor magnetism entered into the same synthesis, contributing to the establishment of a rigorous geophysical model based on the principles of Eulerian geometry.

The observed facts so patiently accumulated by geophysicists over decades ceased to be simply a collection of data and entered the great gate of Science in the Platonic sense of the term, that is, science that can be expressed in mathematical terms. They not only entered, they became one of its most glorious chapters. The observers and measurers found their reward: Nature had finally given up her secrets to them and had repaid them for their long and tedious efforts! Some people consider this a just reward; others scorn it. The systematic work of accumulating observations, without a definite theoretical goal, found a posteriori justification in plate tectonics. Trailing magnetometers behind ships to construct the magnetic map of the ocean, patiently determining the location of earthquakes and their characteristics in order to define their geographic distribution: for years many considered these tasks purely technical and, all things considered, of minor importance. In a climate in which the theoretical is supreme, these laborious tasks often seem ridiculous. In countries like France, where mathematics is the queen of the sciences, such tasks were regarded with contempt: imagine accumulating facts, observations, and measurements without attempting to prove some sort of theory! Nevertheless, the syntheses of Le Pichon and of Isacks, Oliver, and Sykes would not have been possible without these observations.

But the expression *plate tectonics*, pronounced for the first time by Isacks, Oliver, and Sykes, encompassed more than a simple hypothesis offering a framework for seismological or magnetic observations. *Tectonics*, a magic word for geologists, evoked a synthetic explanation for the great crustal movements whose most dramatic manifestation is the building of mountain ranges. Thus, the theory of mobile, rigid plates could be used to provide an explanation for all the geologic and geophysical phenomena taking place on the earth's surface. This lofty goal irritated some and intensely motivated others. Let us follow the progress of plate tectonics as it was used to extend two of its fundamental hypotheses—that plate boundaries are geologically active and that plate interiors are geologically inert—to all geologic phenomena.

Volcanism and Plate Tectonics

Volcanism is the most spectacular of all geologic phenomena, and an object of fascination for photographers and cinematographers as well. To a geologist the fundamental function of a volcano is to bring the molten silicates called magma to the earth's surface. Except for the external core, the interior of the earth is solid. The production of magma therefore implies first of all the melting of part of the mantle. Magma formed in this way must then make a path to the surface. When it succeeds in doing so, it beomes a volcano; when it stops at depth and crystallizes slowly there among the hot rocks, the magma becomes plutonic rock, such as granite. *Magmatic phenomena* are transfers of melted material from the depths toward the surface, from zones of high temperature toward zones of lower temperature. These materials transport heat at the same time. The dual function of volcanism is to transport mass and to transfer thermal energy.

Once the magma arrives at the surface it cools and solidifies into volcanic rock, whose appearance is very characteristic both to the naked eye and to the microscope. Thus it has been possible to detect the presence of volcanic rock in all geologic formations, even the most ancient (3.5 billion years old), and on the surface of the moon. Thanks to photographs taken during space missions, it is known that volcanism has also occurred on Mercury and Mars. Everything seems to indicate that volcanism is common in the solar system and that each time it occurs it implies internal activity in the planet.

The volcanoes presently active on the earth's surface are found in narrow, well-defined zones. Volcanism is most pronounced on the mid-oceanic ridges, although it is very discreet there, since it is mostly submarine. Iceland, which is situated on the Mid-Atlantic Ridge and is subject to intense subareal volcanic activity, is an exception. Dredging of rocks on the oceanic ridges has brought fresh lavas to the surface, indicating abundant volcanism. On the continents emerging ridges, such as the zone of great African lakes, also show strong volcanic activity.

Subduction zones are volcanically active as well. Japan, the Philippines, the Sunda Isles, Chile, Peru, Mexico, and Central

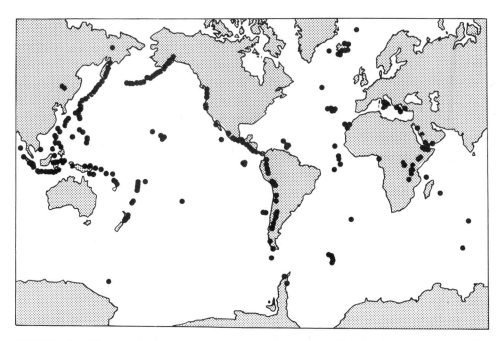

FIGURE 43 The distribution of volcanoes that have erupted during human history. The majority of volcanoes are found near subduction zones or mid-oceanic ridges.

America all contain examples of this kind of volcanism. Only transform faults are volcanically inactive. The well-known problem of the relationship between volcanoes and earthquakes can be clarified in this way: all volcanic areas are seismically active, but seismic areas are not necessarily volcanically active. Nowadays there are no active volcanoes in Turkey, Yugoslavia, or northern Greece, any more than there are near the San Andreas Fault. Only plate edges (whether they are ridges where the mantle creates new oceanic surface or subduction zones where oceanic surface disappears into the depths), where transfers of matter have a vertical component, have volcanoes. Transform faults conserve surface and are the site of horizontal movement only; therefore they lack volcanism. Thus the distribution of volcanism seems to be consistent with the principles of plate tectonics.

Recent Mountain Ranges

Mountain ranges are deformed areas of the earth's surface where the surface has been folded and broken in response to local contractions. Let us look at the arrangement of geologically recent (less than 100 million years old) mountain belts. Using the edges of lithospheric plates as reference points, we can distinguish two types of mountain ranges.

One type of range results from subduction and is located along ocean-continent borders. An example is the American Cordillera, which runs along the western edge of the New World from Alaska to Tierra del Fuego. A series of subduction zones plunges under the continents, except in a section of California that contains the San Andreas Fault, and perpendicular to the trenches lie belts of folded mountains dotted with imposing volcanoes.

The other type is formed by the collision of two continents. An example is the series of mountain belts, called alpine ranges, that separate Laurasia from pieces of Gondwanaland and that extend from the Alps to Thailand, passing through Greece, Iran, the Himalaya, and Burma.

In both cases mountain ranges are forming at the junction of two plates. Thus the distribution of mountain ranges and volcanism reinforces the ideas propounded by seismologists: the earth's geologic activity is confined to plate boundaries.

Plate Tectonics: A Formal System or a Physical Theory?

After the appearance of these seminal works on plates, earth scientists began to assemble all the data, synthesize the various observations and explanations, and formulate a theory of plate tectonics. Here is a historical overview of this attempt to create a rigorous logical system, beginning with the *laws of plate tectonics:*

1. The earth's surface is divided into rigid plates. These spherical sections, about 100 kilometers thick, together form the lithosphere, so the plates are called lithospheric plates.

2. The plates are created at the oceanic ridges. These structures are called accretion zones.

3. The plates move apart from one another without being deformed. They slide on a viscous substrate called the asthenosphere.

4. The plates are destroyed at the oceanic trenches, called subduction zones, where they plunge into the mantle, but only the oceanic parts of the plates are swallowed up in this process.

5. The light continents move with the plates that carry them, but they are not submersible.

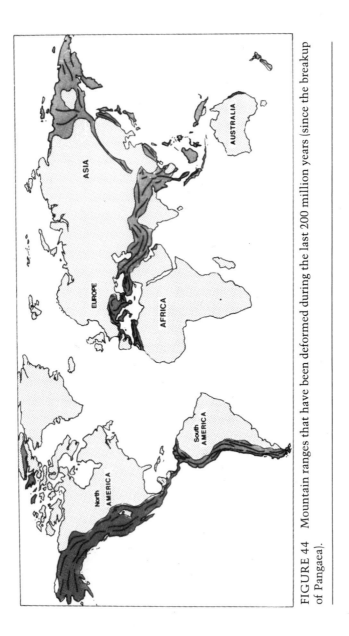

FIGURE 44 Mountain ranges that have been deformed during the last 200 million years (since the breakup of Pangaea).

6. Bordering the plates are ridges (accretion zones), trenches (subduction zones), and transform faults. In general plate boundaries do not coincide with the borders between the oceans and the continents. The study of seismology has made it possible to map the plate boundaries.

7. The energy of the earth's interior is dissipated at the plate boundaries, either mechanically (earthquakes or the formation of mountains) or thermally (volcanoes or the formation of plutonic rock).

8. The relative movements of the rigid plates are governed by the mathematical laws of movement on a sphere. Thus the relative motion of two spherical sections can be completely described in terms of the pole of rotation (Eulerian pole) and the relative angular velocity.

According to the tectonic view, it is possible to reconstruct the geologic history of the earth by defining the ancient plates, the poles of rotation between plates, and the relative angular velocities. These can be used to calculate and reconstruct successive geographies and, in particular, the successive positions of the continental rafts.

In their first global models W. Jason Morgan and Xavier Le Pichon divided the globe into six or seven plates. Data from all areas of the globe—whether they were new observations or measurements or reinterpretations of old information—enriched and complicated this initial plan considerably. Today it is thought that there are about fifteen plates, and new research has produced a surprising complexity and variety of combinations. The relationships among subduction zones, ridges, transform faults, and ocean-continent borders show such varied geometry that it is tempting to develop a different theory to explain each one. In the beginning, however, it was not this variety of natural conditions that attracted interest but, on the contrary, the rigor and formality of the tectonic approach, based on the "theorems" listed above. Supporters of plate tectonics devoted themselves passionately to the determination of Eulerian poles and rates of rotation.

For several years part of the geophysical community was fascinated by geometry on the sphere. Some geophysicists wanted to fit geodynamics into *rigorous theorems* at all costs, in the hope of describing the entire evolution of the earth in terms of a few simple mathematical or topological principles. Transform faults played a fundamental role in this exercise, because they are oriented along a small circle of rotation and can be used to determine the pole of rotation between two plates. Thus, using transform faults and the paths of subduction zones, one can attempt to reconstruct the kinematic history of the earth.

Morgan and Le Pichon constructed a model of the last 50 million years of earth history, but the new theory held the promise of extending beyond this limited paleogeographic framework. Because of its rigorous, quantitative aspect, scientists thought plate tectonics had the potential to predict future plate movements. For example, one could examine an area on a sphere with, say, three plates, a certain distribution of various types of borders, instantaneous poles of rotation, and relative angular velocities and try to calculate how the surface and geometry of each plate would evolve over the course of time.

McKenzie and Morgan tried another type of exercise, which was just as abstract—the study of the evolution of *triple junctions*. A triple junction is a point at which three plate edges meet. There are various types of triple junctions: ridge-ridge-ridge, ridge-fault-trench, trench-trench-fault, and so on. How do the various types of triple junctions evolve? Without denying the inherent interest or the intellectual difficulty of such exercises, I believe they have had very little influence on the evolution of the earth sciences.

Another type of exercise consists in considering the limiting cases in which tectonic principles no longer apply and in which a discontinuity in the system appears. Rene Thom calls such cases *catastrophies*. For example, what happens when a ridge is pulled into a trench? If a ridge stops creating surface, a plate is immediately destroyed and so, in turn, is the adjacent plate. Pursuing the same scenario, one can imagine what happens when two trenches collide. And since the continents are not submersible, the system must become blocked when a continent arrives at a trench. Similarly when two continents collide: both remain on the surface but they are telescoped together in a *continent-continent collision*.

Examining these extreme cases makes us aware of the limits of plate tectonics. When one part of the system becomes blocked, how does compensation of mass take place on a global scale? For example, if subduction is halted as the result of a continent-continent collision, what effect does this have on the distribution of spreading rates, of relative velocities of ridges, or of poles of rotation? When a ridge is engulfed by a trench and disappears into it, what repercussions are felt on the rest of the earth's surface? Is a new ridge born elsewhere? Where? How? Or, on the other hand, are two plates permanently lost? In that case is the earth evolving

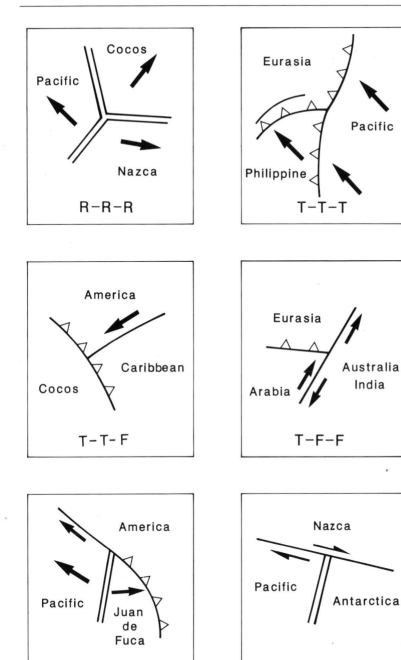

FIGURE 45 Examples of triple-junction configurations. R = ridge; T = trench; F = fault.

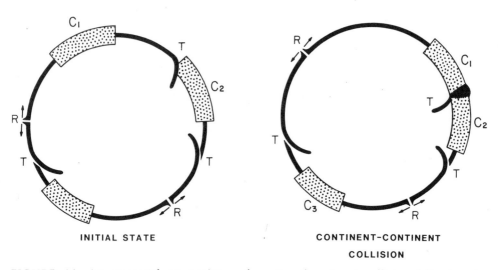

INITIAL STATE

CONTINENT–CONTINENT
COLLISION

FIGURE 46 An equatorial cross-section of continent-continent collision following the complete subduction of the oceanic plate originally between continents C_1 and C_2.

toward having only one or two plates? In that case the earth must have begun with an awful lot of plates.

We know that such behavioral discontinuities do exist. Pangaea, which belonged to a single plate 250 million years ago, has broken up to the point where its fragments are located on six different plates today. But we don't understand how these behavioral discontinuities appear, or where, or why.

J. Tuzo Wilson imagined the following progression of events that would unite the various scenarios into a coherent scheme:

Phase 1: A continent breaks up and a rift valley is formed (the present-day example is East Africa).

Phase 2: Some time later an oceanic ridge is established and a spreading episode begins.

Phase 3: An ocean with a central ridge and two inactive edges (without subduction) is created.

Phase 4: The oceanic floor buries itself beneath one of the edges, creating a new subduction zone (active edge).

Phase 5: This zone later absorbs the ridge itself.

Phase 6: The entire ocean floor is swallowed up in the subduction zone and the two old borders collide with each other,

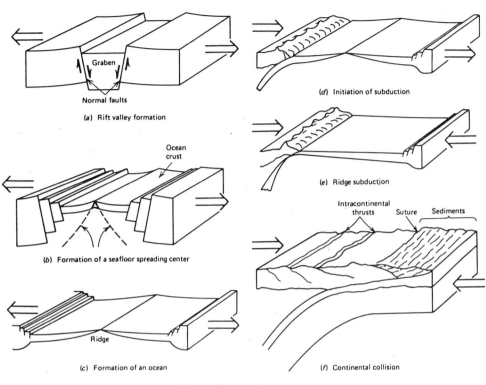

FIGURE 47   Successive stages of the Wilson cycle.

re-forming the continent with a range of mountains where the joint used to be.

When the cycle has been completed in one place, it is repeated somewhere else. According to this projection the geologic history of the earth can be summarized as a series of successive Wilson cycles that are distanced in both time and space among themselves.

With this theory in hand, tectonic geologists believed they had the key to the geologic history of the world.

Reactions against Plate Tectonics

The theory of seafloor spreading had provoked a fairly brief controversy among oceanographers but had left geologists rather indifferent. In their eyes it was an oceanographic theory for oceanographers. On the other hand, plate tectonics provoked a violent reaction within the geological community. Passionate and fanatical debates were held on the subject and, as often happens, the attitudes expressed were extreme and dogmatic to the point of caricature. The violence of the arguments brings to mind the debates between Darwinists and anti-Darwinists at the end of the nineteenth century or, more recently, those over Wegenerian continental drift. A balanced view was not possible: one had to be *for* or *against* plate tectonics. A geophysicist who did not subscribe to the drift theory was considered archaic, but a geologist who showed some sympathy for plate tectonics was immediately classed among the traitors and the adventurers!

Even today I marvel over the suddenness and the violence of these reactions of the early 1970s. Their late date (remember that Robert Dietz's and Harry Hess's articles were published in 1961) may be a result of the amount of time it took for geologists to become aware that the theory of plate tectonics was applicable beyond the ocean shore. Were they prompted by the tectonophysicists' desire to solve such problems of continental geology as the formation of the Alps? Did the other geologists feel threatened by this? The word *tectonics* is used in a specific sense in geology (see Chapter 6); is it not a provocation to use it in an unconventional sense?

It is difficult to answer these questions, but at least we can examine the ways in which this rejection manifested itself and look at the arguments of those who were opposed to the new theory.

Opposing arguments must be put in two categories, which were often mixed together in debates and articles but which were fundamentally different epistemologically: hostility to the idea of continental drift, and more technical criticisms addressed to the rigid rules of plate tectonics or to one of its specific aspects. The first attitude was a global rejection; the second, a disagreement on methods.

The antidrift arguments are, to put it plainly, weak and inconsistent. They result either from a faulty comprehension of the new theories or from a philosophic reasoning far removed from the concrete, scientific facts. For example, one argument holds that because Africa is situated between an expanding Indian ridge and an expanding Atlantic ridge it should show signs of compression. Since it does not—no recent mountain range crosses central Africa—the theory must be false. As I explained in Chapter 3, this argument is in error because it does not take into account that the two ridges are not fixed but are actually moving away from each other.

Other arguments are that the pieces of the North Atlantic cannot be fitted together exactly, that the Pacific cannot be reconstructed, and that the Pacific ridge is not located in the middle of that ocean. Magnetic mapping of the oceans often reveals extremely complex structures: does that not prove that the simple pedagogical forms of anomalies symmetrical to ridges (produced by Vine and Matthews) are only exceptions that lack general applicability? Geologic observation shows that the continents are folded, broken, and deformed at the impetus of young mountain ranges, but also of areas of lower relief that are the remains of old, eroded mountain ranges. If the plates are supposed to be rigid, how can these folds be explained?

Yet more evidence against drift was offered. In the middle of the Pacific plate there are strings of volcanic islands; the best known are the Hawaiian and Tahitian archipelagoes. Absolute dating of the islands shows that, contrary to what J. Tuzo Wilson said, they were not born on the East Pacific Rise but developed in place, in the middle of the Pacific plate. Besides, some of these islands, such as the island of Hawaii, are volcanically active at present. Therefore, it was argued, volcanism is not confined to plate edges; it exists in the very center of the Pacific plate. In an analogous way, there is considerable seismic activity in China, in the very center of the Asian plate. These, again, are violent mechanical manifestations located in the middle of a plate. Plate boundaries thus have no monopoly on discharges of energy!

As is often the case, some of the objections were based on important questions that would have to be answered, but because they were relatively weak none jeopardized the global synthesis offered by plate tectonics. In the natural sciences, where observations are so numerous and the phenomena under consideration are complex, it is absolutely necessary to rank problems according to their importance. Unfortunately, this is not always understood. Thus radical opposition to the idea of lateral movements was rapidly blunted, and the majority of the geologic community was ready to admit to the existence of mobility. But at the same time it was thought that its principles were applicable only to the ocean and did not affect continental geology. Seafloor spreading and plate tectonics seemed to pertain only to the oceans and to be disconnected from the facts of traditional geology. This can be called the theory of the *two geologies,* that of the oceans and that of the continents, with no connection between them. Thus it was possible for some geologists, even very prominent and respected ones, to state that plate tectonics was no concern of theirs.

In summarizing the objections I have simplified them a bit. In fact, the arguments varied among different geologic communities. National cultural contexts play an important part in great ideological debates.

No doubt the American geologic community was one of the most conservative in the world when it came to mobility theory. Antidrift sentiment was rife there. Peter Wyllie wrote that if before 1960 a candidate for a professorship at an American university had shown some sympathy for drift theory, that would have been enough to eliminate his candidacy. But Americans' attraction to new things caused these positions to swing abruptly in the opposite direction under the influence of Robert Dietz, John Dewey, and John Bird. Within two years almost all American geologists became faithful and even zealous partisans of plate tectonics. Today a candidate for a professorship who declared himself against drift would be rapidly dismissed. Obstinate foes such as the Meyerhoffs (Arthur and Howard) were isolated, although protected and venerated, as a tribute to the philosophical liberalism of the New World (just as coelacanths are observed and protected).

In Japan geophysicists rapidly became followers and then convinced participants, but geologists remained totally opposed to the new theory. It wasn't until the petrologist I. Kushiro returned to Japan in 1975 that plate tectonics was taught in the geology department of Tokyo University.

In Europe the struggle was much more protracted, because only a small minority of earth scientists were geophysicists. Since Staub and Kossmat, European geologists had been convinced that some mobility was necessary to explain the Alps; they had a lively tectonic tradition and were totally satisfied to remain within this limiting framework. So they were extremely hostile to the new and grandiose mobility that seemed positively caricatural in its excess. In France plate tectonics was first taught only in 1970, after pedagogical autonomy was granted to the universities and the geology department of the new University of Paris-VII was created from scratch, with no encumbering heritage. In 1971, ten years after Hess's article, the only leading geologist to accept continental drift, even as a hypothesis, was Maurice Mattauer of Montpellier. In 1978, ten years after the work of McKenzie, Parker, Morgan, and Le Pichon had appeared, more than half of the French geologists at a national congress in Orsay were still hostile to plate tectonics! Everywhere the conflicts took a sharp and violent form and the scientific community split into factions. For more than ten years the scientific page of the newspaper *Le Monde* was a richer source of information on plate tectonics than texts for high schools or universities!

Other European countries were not much better off, except for Great Britain, thanks to the penetration of American articles and to the geophysics departments at Cambridge (in particular Teddy Bullard and his students Drummond Matthews, Fred Vine, Dan McKenzie, and John Sclater) and at Newcastle (where Keith Runcorn and Dave Tözer taught). In general, British geologists were hardly more receptive than their European colleagues.

In this roll call of Philistinism the Soviet Union gets highest honors for its static and reactionary attitude. Research, which took place mostly in the laboratories of the Academy of Sciences, was separated completely from teaching and from the intellectual debate of students. The Academician Vladimir Belousov, grand old man of Soviet geology, was an obstinate foe of mobility, and the entire Soviet geologic community remained hostile to drift theory into the 1980s. Then, finally, young and courageous scientists undertook a patient reconquest of minds! Thus a great scientific country with a rich tradition in the earth sciences and many talented scientists remained outside this revolution in thought for more than ten years.

But in the end all opposition to mobility theory was overcome or set aside. Fifty years earlier Wegener had been almost alone against a hostile geologic community. In the confrontation of the 1970s the conflict was a case of two opposing communities. The promobility community was young, active, and numerous, and students rallied to it en masse. Like a breaking wave, the new theory swept away all other ideas.

Today continental drift is as accepted a scientific fact as the structure of the atom or the chemical formula of DNA. All geologic research must be put into the framework of mobility theory, and no one contests the idea any longer. No doubt this framework is more flexible than the one developed by the theoreticians of plate tectonics, but fundamentally it is the same as that defined by Hess and by Isacks, Oliver, and Sykes. This approach has generated a revolution in the earth sciences that has followed two complementary paths: on one hand, it created a new method for studying the earth in terms of what a given region represents in the scenario of plate tectonics (I call this approach "plate geology"); on the other hand, it transformed the traditional disciplines of geology profoundly.

Some people call this new way of approaching the earth sciences *global geology.* It is global both in its scope and in its multidisciplinary approach. As in Wegener's time, researchers in the various disciplines of the earth sciences are all attempting to understand the functioning of a single object of common study: the earth. No doubt plate tectonics' role in unifying the geological sciences has become just as important as the theory itself.

THE BIRTH OF MARINE GEOLOGY

5

BEFORE the new theory of plate tectonics was accepted a geology of the seafloor certainly existed, but it amounted to only a collection of measurements, maps, and samples; no links or causal relations among them were agreed upon. Mobilist ideas provided marine geology with both foundations and extremely powerful tools for analysis, one of the most important of which was magnetic cartography. A new chapter in a global geology capable of exploring the hitherto unknown two-thirds of the earth's surface now opened. Let us look at the foundations and the first steps in this process, which is still in progress.

The Crumpled Zebra Skin and the History of the Oceans

The ocean floor is composed of basalt covered with sediments. The basalts retain a magnetic signature that holds two messages: one is *chronological*, making it possible to date the various zones of the ocean floor by reference to the scale of magnetic reversals, and the other is *geometrical*. Because the magnetic anomalies that are recorded in basalts as they are formed run parallel to the ridges, they are parallel to each other. If the parallelism is destroyed, a change in the conditions of seafloor spreading is indicated. It can be a change of orientation or of rate, differential sliding along a transform fault, or the birth or disappearance of a ridge. The zebra-skin pattern of the map of the positive and negative anomalies is the memory of the oceans' geologic history.

To reconstruct the history of an ocean one proceeds step by step. After having identified and numbered the magnetic anomalies on both sides of a ridge, one superimposes anomalies of the same age; this "closes" the seafloor between them and allows one to see a picture of the ridge when those magnetic anomalies were created. Through repeated trials of this procedure the history of the oceans of the world was reconstructed. These methods had already been used, but after plate tectonics we find them employed in an even

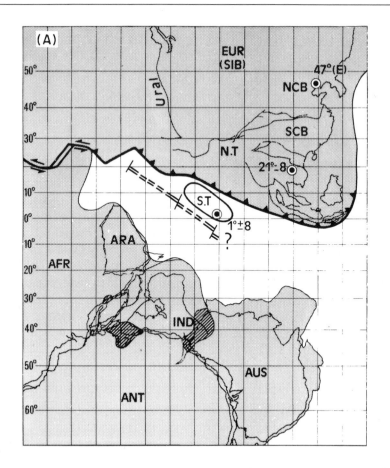

FIGURE 48 "Snapshots" of the northward drift of India and associated development of spreading centers in the Indian Ocean.

(A) 150 M.Y.B.P. India, Africa, Asia, Antarctica, Australia, and Arabia are still united as the Gondwana supercontinent.

more precise manner. Xavier Le Pichon had shown the usefulness of the magnetic technique for a relatively simple global case. He was followed in deciphering more complex cases by John Sclater, Dan McKenzie, Roland Schlich, and Philippe Patriat, for the Indian Ocean, and Walter Pitman and Tanya Atwater, for the East Pacific. When compared with the basic outline, these studies showed the existence of complicating factors.

The Indian Ocean, for example, is almost completely enclosed today. It is surrounded by an almost continuous continental

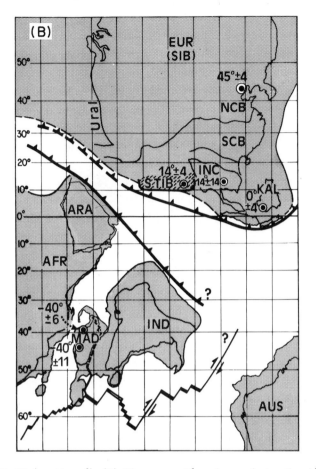

FIGURE 48 (*continued*) (B) 83 M.Y.B.P. The circum-Antarctic ridge has formed and Africa-India and Antarctica-Australia have separated. Now a new ridge propagates northward from the circum-Antarctic ridge and divides India from Madagascar. (Madagascar-India had already separated from Africa during an earlier episode of northward ridge propagation.)

mass—Africa on the west, Asia to the north (represented by the prow of India and the Sunda Isles), Australia to the east, and Antarctica to the south. Its topography is dominated by the star-shaped arrangement of the three ridges that fan out from the Rodriguez Triple Junction. In addition to the special appeal of a ridge-ridge-ridge triple junction predicted by the plate-tectonic theory, the Indian Ocean attracted study because the three

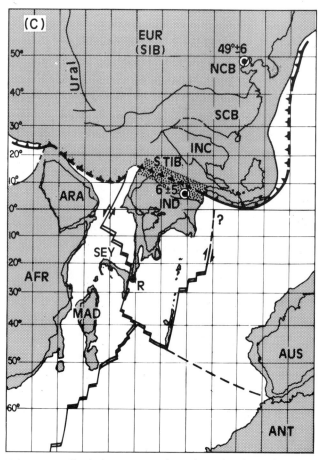

FIGURE 48 (*continued*) (*C*) 48 M.Y.B.P. The Indian plate has subducted to the point where the Indian and Asian continents collide. The complex geometry of the Indian-Asian subduction zone results in not one collision but a succession of collisions. (Imagine in the future the complicated set of collisions that will result if Australia–New Guinea continues to move northward and collides with the Philippines and then Asia.) The collision of India and Asia is also associated with a major reorganization of spreading centers in the eastern Indian Ocean, which led to the propagation of spreading between Antarctica and Australia.

branches have very different spreading rates (5 centimeters per year for the north ridge, 2 centimeters per year for the southwest branch, and 10 centimeters per year for the southeast).

Like the South Atlantic, the Indian Ocean resulted from the breakup of Pangaea. Although the global scenario had been known since Wegener's time, the exact mechanisms for explaining the

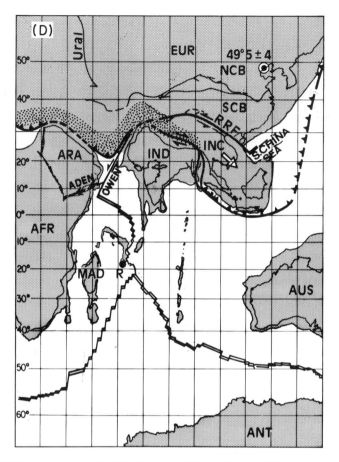

FIGURE 48 (*continued*) (*D*) 23 M.Y.B.P. The collision of India with Asia by this time is resulting in substantial Asian deformation, leading to the formation of the South China Sea. The Owen transform fault connecting the Central Indian Ridge to a greater Himalayan subduction/convergence zone is replaced by a western propagation of the Central Indian Ridge. This results in the separation of Arabia from Africa and the formation of the Red Sea spreading system.

ocean floor's morphology or the magnetic map of the Indian Ocean became known only as a result of modern studies directly inspired by plate tectonics. Today it is clear that India "crossed" the Indian Ocean to attach itself to Asia, but this process was not straightforward. Fifty-five M.Y.B.P. the Indian ridge was oriented east-west. A gigantic north-south transform fault (the Chagos Fault) separated the two parts of the ocean and India began to drift northward.

Then, 35 M.Y.B.P., the Chagos Fault disintegrated into a complex structure consisting of a series of small ridges cut by transform faults. The whole structure amounted to a staircase of ridges with a northwest-southeast orientation. At this point India began to change direction and drift toward the northeast. The breakup of the Indian ridge continued until it reached the degree of complexity that it exhibits today (see Figure 48). Moreover, the seafloor spreading rates have varied. Fifty-five M.Y.B.P. they were very fast (16 centimeters per year); since then they have slowed down abruptly several times to the present rate of 5 centimeters per year. Further complications arise because the Indian Ocean contains structures that do not relate very easily to the simple theory of seafloor spreading, such as the Ninety East Ridge, which is oriented north-south and is aseismic, and the Mascarene Basin, with its appendix, the granitic Seychelle Islands. Where do these structures come from? How did pieces of continent (the Seychelles) become isolated in this way? How were the volcanic structures (Ninety East Ridge) created? Plate tectonics does not answer these questions, but at least it allows us to formulate them!

Another interesting question involves the end of the north branch of the Indian ridge. The east and west branches of the ridge are linked to the worldwide network of ridges, but the north branch disappears into the Red Sea. Does it rejoin another plate edge? What is its relationship with the great African rifts? These questions are less exotic than the preceding ones and we will be able to answer parts of them.

The Indian Ocean is not unique in its complicated history and morphology. The structure of the eastern region of the Pacific, the area that adjoins the two Americas, is dominated by the great ridge called the East Pacific Rise, which has posed quite a challenge to geologists. Bill Menard and Tanya Atwater, then working at Scripps, studied the magnetic anomalies of the west coasts of North and South America. They made a detailed reconstruction of the movements of the plates as recorded in the magnetic anomalies and showed that the oceanic ridges change direction and shape. The magnetic structure of the ocean floor records these continual changes. Walter Pitman and Jim Hays of Lamont, studying the anomalies in the Gulf of Alaska, also showed the changes in direction, as well as the migration of the ridge in relation to the coast of the continent. Adjacent to the continent are two subduction zones, the Alaskan zone and the British Columbian zone, which serve as a model of the limiting case in which the ridge arrives at a subduction zone, is engulfed

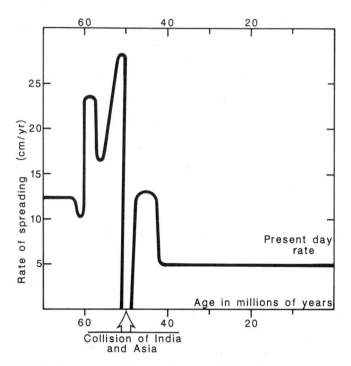

FIGURE 49 Variations in the spreading rate of the Indian Ocean in the past 70 million years. The major changes in spreading rate are more or less associated with collisions in Asia. This curve was determined from seafloor magnetic data and paleomagnetic data from Asia. (Based on the work of Jose Achache and Philippe Patriat.)

by it, and stops spreading. This phenomenon—a surface-creating plate boundary being eliminated with a single stroke—has important consequences.

In fact, interaction between a ridge and a subduction zone takes place all along the west coast of North America. Tanya Atwater's detailed study of this interaction and of the evolution of the East Pacific Rise led her to undertake a remarkable reconstruction of the geologic history of the western American region for the past 100 million years. For this purpose she introduced a new plate, the Farallon, which has practically disappeared today. She explained the birth of the San Andreas Fault as well as the chronology of volcanism in the American west by the swallowing up of Farallon. Generally speaking, these studies show that engulfing a ridge stops the process of subduction and tends to transform a subduction zone into a transform fault. Is this a general phenomenon or a

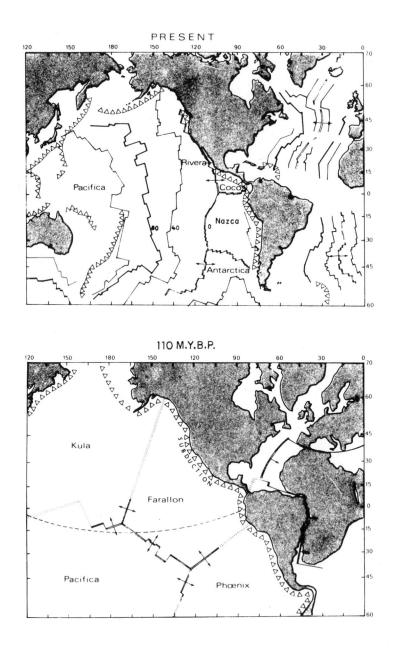

FIGURE 50 *Top:* Present locations of mid-oceanic spreading centers and magnetic anomalies created 40–80 M.Y.B.P. *Bottom:* The situation 110 M.Y.B.P. Subduction at trenches along the west coast of the Americas has since led to the disappearance of the Farallon and Phoenix plates.

particular instance? That is a fundamental question for plate-tectonic theory, for which we still do not have a general answer.

"Relativistic" Stratigraphy

Stratigraphy is the study of layers of sediments (strata) lying one on top of another. The layers are the successive pages of the book recounting the geologic history of a place. The pages are read from bottom to top, from the oldest layers to the youngest. That, at least, was the classical concept. In a growing ocean, the situation is more complicated.

The morphology of the seafloor is the starting point for any ocean's history. The ridges are zones of thermal swelling, and a shallow seafloor indicates that hot mantle lies beneath it. As it drifts away from the ridge on both sides, the floor cools off, contracts, and therefore deepens. In other words, as one moves away from the ridge, the depth of the ocean increases according to a quasi-exponential law.

The laying down of sedimentary strata in the oceans is governed by two factors: distance from the continents and biological activity. The continents furnish the products of erosion, the large and small solid particles. Once delivered to the ocean, these particles fall to the bottom. The larger they are, the faster they fall. Therefore they can be grouped by distance from shore: the largest near the coasts, the smallest far out in the open sea; the pebbles and the sands on the beaches, the clays in the great depths.

Sediment of biochemical origin is added to these products of mechanical erosion. Marine-biological activity is governed entirely by the plankton that furnishes the basic food for the marine animal species (that is, zooplankton, invertebrates, and fishes). Most of these species form shells, or tests, that contain either calcium or silicon. When they die, their tests fall to the bottom, accumulating as calcareous or siliceous sediments. The biological activity of shell-bearing organisms is greatest in the warm equatorial zone, where the luxuriant marine life produces a great many calcareous tests. In cold water the tests formed are mostly siliceous, rather than calcareous, and overall biological activity is lower. Sediments from the biological tests are therefore characterized by zones: calcareous near the equator, more siliceous nearer the poles.

The equator is marked by a calcareous band on the map of sedimentation. These sedimentological keys, added to the ideas suggested by mobility theory, helped to unlock the messages in oceanic sediments. But sedimentation reflects this zonality in an

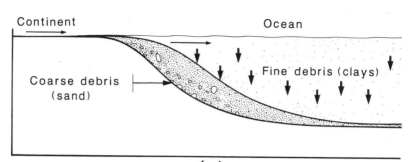

(a)

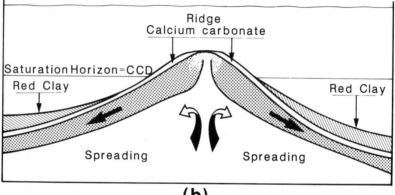

(b)

FIGURE 51 Seafloor sedimentation is strongly influenced by three factors: the distance of the seafloor from a continent, its location in the tropical region of high biological productivity versus location in less active regions nearer the poles, and the depth of the seafloor. (*a*) Coarser sediments washed off continents tend to settle out near the continent. Finer sediments are transported farther away from the continent before they settle out. (*b*) Calcium carbonate becomes more soluble in sea water as depth (and pressure) increases. Thus in shallower waters near mid-oceanic ridges calcareous oozes accumulate faster than they are dissolved. In deeper water, below the carbonate compensation depth (CCD), calcareous oozes dissolve faster than they are deposited and only red clays can accumulate.

imperfect way, because another phenomenon, redissolution, is superimposed on it. Calcium carbonate is very soluble under pressure and thus dissolves in the great marine depths. The calcareous tests are formed near the surface. Light from the sun, the source of photosynthesis and therefore of planktonic activity, does not penetrate very deep, so biological activity is concentrated in the first 200 meters of surface water. After the death of the animal, its test falls toward the bottom. If the bottom is deep, the test is redissolved before reaching it. As a result, sediments are calcareous only at shallow depths; in deep water only clays are laid down. And because the ocean floor is a conveyor belt, the sediments laid down at the axis of an oceanic ridge are displaced several hundred kilometers away from it after a few million years. Away from the ridge more sediments are laid down; they in turn also drift away, and so forth.

Where there is no subduction zone between a ridge and a continent (in which case the transition between seafloor and continental crust is called an *inactive continental margin*), the sedimentary column is rather homogeneous from bottom to top at every point in the ocean near the continent. The sediments come primarily from the continent. Farther out the sedimentary column is a bit more complicated, but it remains almost constant over the course of time, because sedimentation depends on the distance from the continent, which does not vary over time.

On the other hand, where there is an active margin—a subduction zone—between the ridge and the continent, the seafloor approaches the continent with time. In this case the interpretation of the sedimentary column is more difficult. Such are the foundations of the new chapter of geology called marine stratigraphy. To practice it, we need samples, in proper order, and to obtain samples we must drill through the ocean floor under several kilometers of water. This technically difficult task necessitated the undertaking of a gigantic program of underwater drilling, the Joint Oceanographic Institutions for Deep Earth Sampling. JOIDES was an American project in the beginning but was later extended—and absorbed sums of money amounting to 35 million dollars per year—as the International Phase of Ocean Drilling (IPOD) project. It mobilized a scientific community that was reticent and few in number at first, then more and more enthusiastic and numerous. The data it collected systematically in all the great oceans of the world formed the basis for the new oceanic stratigraphy.

The History of the Atlantic

Except for trenches near the Antilles and the Sandwich Islands, the Atlantic is bordered only by inactive continental margins. For this reason Atlantic stratigraphy is rather simple.

The dating of the first sediments dredged from the Atlantic verified that the North Atlantic and the South Atlantic did not open up at the same time. This had been shown by the magnetic anomalies but marine stratigraphy, with its dating by means of microfossils, confirmed it and made it more precise. Laurasia broke up before Gondwanaland did, and North America broke apart from Africa in the mid-Jurassic (165 M.Y.B.P.). The birth of the South Atlantic did not take place until 125 M.Y.B.P. The separation of North America from Europe was completed by the opening of the northern part of the Atlantic 80 M.Y.B.P.

Stratigraphic methods also made it possible to describe the successive geographies of the Atlantic region. The fragmentation began with phenomena comparable to those found today in the African Rift Valley near Ethiopia and the Afar Triangle. Volcanism, crustal extension, and the birth of salt lakes are its permanent manifestations. In shallow water near the nascent Atlantic continental margins, saline deposits and deposits rich in organic matter were laid down; these would later be compressed to form the oil fields that lie along the coast of the Atlantic, as in Venezuela and Nigeria. The increasing width of the ocean, the mechanism of seafloor spreading, and the evolution of the topography of the ocean floor under the influence of thermal contraction all led to the laying down of calcareous sediments near the ridge and of sediments rich in continental debris, such as sand, near shore. When the Atlantic had grown wide enough, fine sediments filtered down to the seafloor far from the coasts as clays.

The History of the Pacific

The history of the eastern Pacific was deciphered from the record of magnetic anomalies, but this technique is difficult to use in the central and western Pacific because much of the seafloor was formed in the Cretaceous or lower Jurassic periods, when there were no reversals of the magnetic field. (When the direction of magnetization remains uniform the period is said to be "magnetically quiet.") In this case the floor cannot be "oriented" on the chronological scale of magnetic reversals (see Figure 19). The history of this part of the ocean floor, then, had to be discovered through stratigraphic methods.

Further complications arise because the Pacific is surrounded on all sides by subduction zones. Drilling showed a much more complex geologic history than that of the Atlantic. When cores were drilled along a line moving away from the ridge, they at first showed the expected configuration: calcareous sediments topped by clays. The conveyor belt, which carries the sediments, receives calcareous sedimentation near the ridge; as it cools, the floor

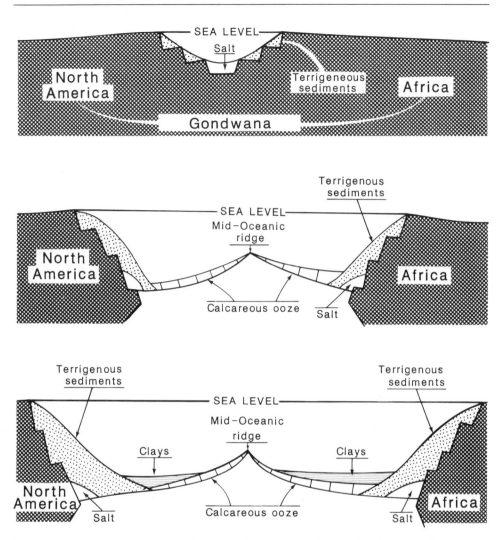

FIGURE 52 History of the opening and sedimentation of the Atlantic Ocean. In the early stage, shallow deposits rich in salt and organic matter are laid down with volcanic debris. In the small sedimentary basins of the second stage, only sand and coarse detritus are deposited because the basins are so close to the continent. Finer sedimentation, such as clays, are formed in the abyssal plain of an open ocean. Note that salt and organic sediments laid down in the early stage are buried under thick sand and clay deposits.

receives clays and sinks as it moves away from the ridge. In regions that are distant from the equatorial zone today, however, extensive calcareous deposits are sandwiched between clays. Did a secondary volcanic ridge decrease the bottom depth? There is no volcanic or topographic evidence of such a thing. After various unfruitful ideas

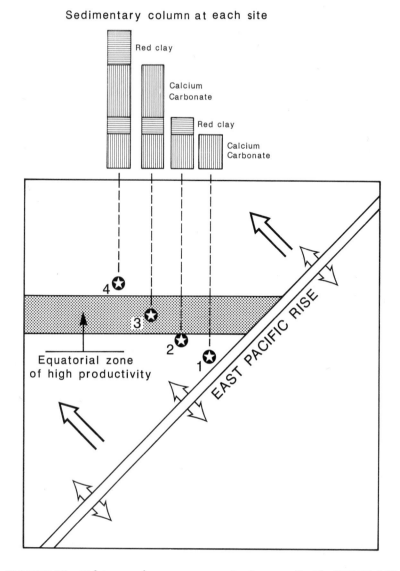

Sedimentary column at each site

Red clay

Calcium
Carbonate

Red clay

Calcium
Carbonate

Equatorial zone
of high productivity

EAST PACIFIC RISE

FIGURE 53 Calcium carbonate occurs twice in some Pacific JOIDES drill cores. (1) Carbonate is deposited near the East Pacific Rise, where the seafloor is shallower than the carbonate compensation depth (CCD). (2) As the seafloor moves away from the ridge it becomes deeper than the CCD, so only red clay is deposited. (3) As it moves northward, however, the seafloor also moves into equatorial regions of high biological productivity, where the CCD is deeper than in less productive regions. In this region calcium carbonate is again deposited. (4) Finally the seafloor moves north of the equatorial high-productivity region to a region where the CCD is again shallower than the seafloor depth. In this region only red clays are deposited.

were tested, the hypothesis has been accepted that a change in the direction of seafloor speading is responsible for the unusual sedimentary sequence. The Pacific seafloor moved northwest 40 M.Y.B.P. At that time the conveyor belt crossed the biologically fertile equatorial zone, which explains the recurrence of calcareous sediments. For the past 35 million years the direction of spreading has been westward; because the conveyor belt no longer cut across the biologically fertile zone, there were no more calcareous sandwiches in the stratigraphic records.

The works of Pitman and Atwater brought to light small-scale changes in the spreading direction. Marine stratigraphy, on the other hand, uncovered a change of spreading direction on the scale of a whole ocean, and what an ocean—it is almost 10,000 kilometers wide! Furthermore, stratigraphic methods made it possible to detect and then reconstruct this movement. (The reconstruction based on the study of sediments was corroborated by the study of volcanic ranges, which I will mention in reference to the theory of plumes in Chapter 8.)

All these reconstructions of the history of the world's great oceans, which opened up opportunities for the exploitation of the mineral and petroleum riches that they contain, are a direct product of plate tectonics. Oceanic geology is in full bloom, but these phenomenological studies are not yet complete.

6   PLATE BOUNDARIES

THE DEVELOPMENT of plate tectonics inspired geologists to study plate boundaries, the regions in which geologic activity takes place. *Plate geology* is first of all the geology of plate boundaries: a geology of ridges, where the lithosphere is created; a geology of transform faults, where evidence is seen of the sliding of the spherical sections; a geology of subduction zones, where the plates disappear into the mantle. In the tectonic system each border is well defined, but the role it plays is not made explicit by the theory. The object of studying plate boundaries is to define these roles.

Oceanic Ridges

A ridge is first of all an elongated topographic swell, rising above the submarine abyssal plain to a height of nearly 3,000 meters. Imagine a mountain range crossing a plain and separating it into two provinces: that is how the Atlantic would look if it were possible to dry it up. These mountains form a network that crosses most of the world's oceans. But, unlike the Alps or the Appalachians, the ridges are volcanic in nature. An oceanic ridge is a series of volcanic ranges through which lava flows and in which volcanic cones, calderas, and steep escarpments alternate.

What is the relationship between this volcanic geologic structure, as we perceive it, and its function as creator of lithosphere, as defined by plate tectonics?

The Internal Structure of Oceanic Plates

By taking seismic measurements at sea scientists, mainly at Lamont and Scripps, determined the structure of the ocean floor. They detected below the sedimentary layer (layer 1) three layers of rock (layers 2, 3, and 4) having different seismic-wave velocities.

Laboratory experiments in which the sound velocities of different rocks are measured by ultrasonic devices made it possible to create a test for rocks dredged up at sea. The researchers wanted to know which rocks matched the seismic velocity measured for each layer by seismologists. After some debate the marine geophysicists agreed that layer 2 is certainly made of basalt; layer 3, of a kind of coarse-grained basalt called gabbro; and layer 4, of a peridotite. The transition between layer 2 and layer 3 is the Moho, the limit between the crust and the mantle. Therefore, as expected, oceanic mantle and continental mantle have the same composition: peridotite.

The oceanic lithosphere has a simple structure: 6 kilometers of basaltic rock above 65 kilometers of peridotite. This two-layer structure was confirmed and refined through a series of indirect studies of very special assemblages of rocks, called *ophiolitic associations*, that can be observed in cross-section on the continents.

Ophiolites

While studies were being undertaken to determine the structure of oceanic ridges, some geologists were becoming interested in rocks that, although they were found on the continents, had an oceanic "flavor." These groups of rocks, known as ophiolites, played an essential role in the development of the "new geology."

Ophiolitic associations had long been known to alpine geologists, who in their structural description of "tectonic units displaced horizontally," or overturned folds, recognized layers composed almost completely of ophiolitic rocks. The ophiolitic sequence consists of three parts. Peridotites, most often altered and transformed into serpentines, are found on the bottom, gabbros in the middle, and basalts (called pillow lavas) on top. On top of the basalts are sedimentary rocks containing siliceous fossil protozoans called radiolarians.

Ophiolitic associations are usually deformed, broken up. In fact, the reconstruction of a regular ophiolitic sequence is a delicate operation, because tectonic movements always fragment these suites into separate and scattered pieces in the middle of gigantic overturned folds. The existence of basaltic pillow lavas suggests submarine discharges. That is why starting in 1950 European geologists had suggested a relationship between ophiolites and oceanic rock; because they had not worked out the consequences of this relationship, however, their interpretation remained anecdotal. Besides, these rocks seemed so specific to the alpine ranges that their more general significance was difficult to guess.

The acceptance of plate tectonics marked a revival of interest in rocks. Various geologists in different parts of the world suggested almost at the same time that ophiolites are pieces of oceanic crust that have been transported onto the continents during orogenic phenomena (that is, during the formation of mountain belts). This idea was proposed by Ian Gass of the Open University in England, Eldridge Moores of the University of California, Bob Coleman of the U.S. Geological Survey, Joe Cann, then at the University of East Anglia, Hugh Davies in Papua New Guinea, and John Dewey and Jack Bird, whose efforts in propagating tectonic ideas among geologists will be discussed later. Through their work it was learned that ophiolitic associations are not limited to alpine ranges. To the alpine ophiolites of Cyprus, Greece, and Turkey were soon added those discovered in Papua New Guinea, New Caledonia, Newfoundland, the western United States, Guatemala, and more—briefly, in varied geographic locations but always linked to mountain ranges and, within these ranges, intimately associated with great internal foldings and great lateral drifts. Because the ophiolites' origin is the ocean floor, they provide natural cross-sections of the oceanic crust and of a part of the lithosphere in easily accessible places (the continents). Therefore the model developed from observations at sea can be tested and improved. It is easy to see how interesting it was for those who were trying to understand ridges to be able to study ophiolites.

Some researchers, such as Joe Cann and Bob Coleman, looked at ophiolitic associations from the petrological and structural viewpoint; others, such as Fred Vine, looked unsuccessfully for magnetic anomalies in them or, more simply, for proof of the existence of inversions. The group that I head in Paris concentrated on their geochemical aspects. The various criteria for comparison (mineralogical, chemical, or isotopic composition) all led to the same conclusion: the rocks of the ophiolitic sequence surely belong to the same "family" as the rocks dredged up from the oceans. An ophiolitic basalt resembles an oceanic basalt; an ophiolitic gabbro, a gabbro dredged from the ocean; an ophiolitic peridotite, an oceanic peridotite. The resemblance extends to and includes the extreme serpentinization of the two rocks. Studying the ophiolitic sequences confirmed the geologists' two-layer model: a basaltic crust (true basalts on top, gabbros beneath) and a layer of lithosphere, composed of peridotite. To construct a model for the oceanic ridges and to understand how the oceanic lithosphere is

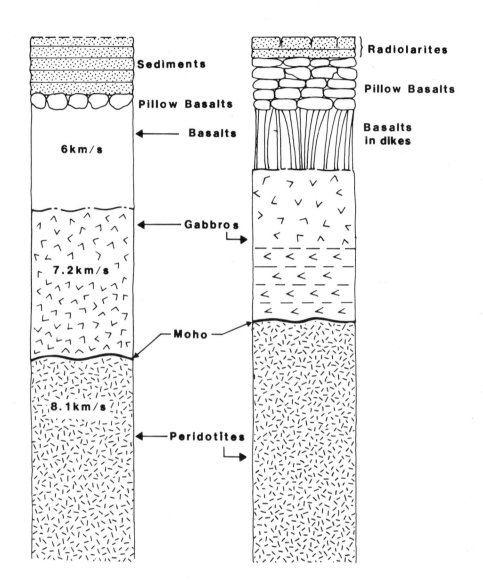

FIGURE 54   *Left:* Cross-section of the layers of the oceanic crust as determined by seismological investigations. The numbers give velocities for seismic *P*-wave propagation in the various layers. *Right:* The layers of the oceanic crust as determined by field observations on ophiolites. By measuring the seismic velocities of the various sections of an ophiolite we can correlate each layer of an ophiolite with a seismically mapped velocity layer and thus arrive at a chemical interpretation of the seismic cross-section.

made, they searched for a mechanism to explain the genesis of these two layers: basaltic crust and peridotitic lithosphere.

The Dynamic Ridge Model

The melting of a mixture of several minerals of different compositions, from which comes a rock like peridotite, is not a simple matter. A pure substance has a *melting point,* but a mixture melts in steps through a *melting interval.* Each step is marked by the melting of a particular mineralogical assemblage. Thus it is possible to produce by partial melting liquids that do not have the same chemical composition as the solid with which one began. For example, the result of melting 10 to 20 percent of a peridotite is a basaltic liquid.

Pressure also has an effect on melting. The melting of an object corresponds to the partial disorganization of the atoms that compose it. Heat provides the molecular energy of atoms and therefore promotes their independence; when this energy is sufficient, the atoms acquire a degree of independence corresponding to what is called the liquid state, the temperature then being called the melting temperature. Pressure, on the other hand, keeps the atoms together and resists their independence. When the pressure is high, more energy must be furnished to liberate the atoms and allow them to pass into the liquid state. That means that the melting temperature under high pressure is higher than that under ordinary pressure. On the other hand, if the temperature is close to the melting point under pressure and that pressure is removed, the atoms take on a certain independence and the substance begins to melt.

In geology pressure and depth go together. The deeper one goes, the higher the pressure; the higher one goes, the lower the pressure. Let us apply this principle to a piece of peridotitic mantle that is suddenly brought to the surface. The swiftness of the trip prevents it from cooling off by losing heat along its edges. If the temperature was not far below its melting point at depth, it melts spontaneously as a result of decompression. Since peridotite is a mixture of minerals, the principle of complex melting just described applies and a basaltic liquid results. The piece of mantle rising toward the surface, which was initially in the solid state, is transformed into a "sponge" full of liquid. The proportion of liquid increases during the rise. Suppose that near the surface this sponge is squeezed: liquid will be ejected upward, and the residual solid—what remains after the extraction of the liquid—will settle to the bottom.

Jan Bottinga and I proposed this mechanism of melting, perco-

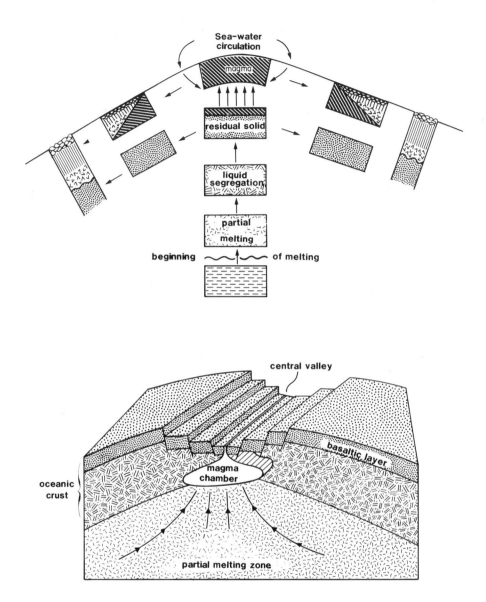

FIGURE 55 The creation oceanic lithosphere. Material from the mantle rises beneath a ridge axis. During its ascent melting occurs and the liquid thus formed rises faster than surrounding unmelted material. Near the surface the liquid (which forms the crust) cools and interacts with seawater. Then the liquid and residual material spread horizontally away from the spreading axis. On the bottom is a block diagram of a slow-spreading ridge (spreading at, say, a rate on the order of 2 centimeters per year) that shows how the processes of lithosphere creation may occur in a more realistic scenario.

lation, and segregation to explain the formation of oceanic crust at the ridges. Our model renewed the first dynamic picture of the ridges that Ron Oxburgh and Don Turcotte of Oxford had proposed, and it integrated both Joe Cann's qualitative observations and geophysical measurements in a quantitative model. The residual "sponge" corresponds to those peridotites that are found under the gabbros in ophiolites and that, in fact, have the chemical character of residues. The chemical elements that enter preferentially into the composition of basalts are absent from these residual perido-tites. The expulsion of a basaltic liquid is the origin of volcanism and of the pillow lavas that are so often found on the ocean floor. Under the roof formed by the piling up of these lavas, a liquid remains imprisoned. A cavity full of this trapped liquid is called a magma chamber. The magma cools slowly, allowing crystals to form. Because cooling is slow, the crystals grow large and form large-grained basalts called gabbros.

The ascent of the mantle and the melting and expulsion of the magma is conceived as a continuous process. Once formed, each new piece of lithosphere drifts laterally; it is immediately replaced at the axis by a new piece of lithosphere, and so forth. This physical model of ridge processes fits neatly into the classic tenet of plate tectonics, namely, that the ridge continuously makes new pieces of lithosphere in an extremely narrow central region, with a constant rate of production over time for a given region.

But this is only a model. It must be tested against the facts, that is, against detailed observations of the ridges.

The Exploration of Emerged Ridges

Iceland is the one place on earth where an oceanic ridge breaks the sea surface. An island of 103,000 square kilometers, it has an essentially volcanic structure. Because its northern location and history limit its vegetation to a few scrubby plants and bushes, its structure can be observed easily: it is not, as tropical countries are, hidden by dense vegetation. The countryside is uniformly volcanic, as far as the eye can see; layers of lava flows reach depths of hundreds or even thousands of meters. In the central region these flows alternate with real volcanoes with cones, volcanic slag, and lava built up to imposing heights. In the north, not far from the city of Akureyri in the area of Lake Mývatn, clusters of cones of various dimensions can be seen. A close study of this area reveals that the cones are aligned along fissures. In the south, nearer to Reykjavik, there are extremely active volcanoes such as Hekla or Heimaey in the Vestmannaeyjar islands. The permanent glaciers in the center of the island give birth to very characteristic volcanic formations: subglacial volcanoes. The lava flows there take the form of pillows, as they do under the sea. The majority of these lavas are basaltic in nature.

Detailed mapping by Icelandic geologists, who are few in number but very dynamic, showed that the active volcanism of the present day is concentrated in an east-west strip from the Reykjanes Ridge in the North Atlantic (a ridge on which "classic" magnetic anomalies can be seen) to the central Vatnajökull glacier, and a north-south strip from the Mývatn zone to the Hekla volcano or even to the island of Surtsey toward the south. The rest of the island exhibits older volcanism.

The active zone consists of a central valley bounded by normal faults typical of extension zones. A recent study was able to show in a more precise way that there is communication between volcanic cones and lava flows from fissures. The volcano seems to fuel lateral injections that sometimes spill out over the surface. In

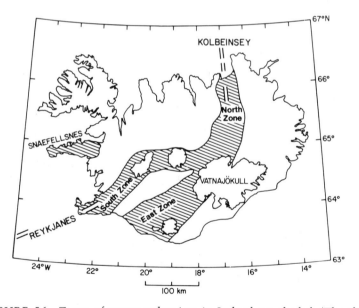

FIGURE 56   Zones of recent volcanism in Iceland are shaded. (The three white spots show permanent glaciers.) The southern zone continues into the ocean as the Reykjanes Ridge and the extension of the northern zone is the Kolbeinsey Ridge. The eastern zone marks a former zone of spreading or a failed rift.

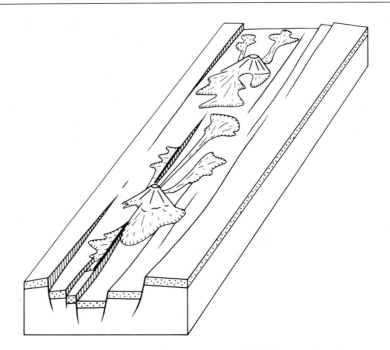

FIGURE 57 The association between normal faulting and volcanism in Iceland. Recent volcanism is shown in relief around the volcanic cones.

other words, the volcano is the supplier and the lavas are the overflow spilled out on the sides. Such activity is observed near the volcano Krafla.

It is not possible to find in the central valley a central fissure through which lava rises continuously and separates the sides, as would be expected in a naive version of tectonics. The relationships between the volcanic apparatus and the lava flows are extremely complex; each new volcanic structure appears more or less at random. The central valley, which is 10 kilometers wide, appears to contain the plate boundaries. It is much more difficult to determine their exact location within this zone.

Outside the central valley the mountains are higher and the rocks are older, all of which is consistent with the idea of the spreading of the ocean floor.

The Exploration of Submerged Ridges

Only with the development of modern technology, particularly submersibles, did it become possible to study oceanic ridges

Era	Period	Epoch	M.Y.B.P.
Cenozoic	Quaternary		
	Tertiary	Pliocene	12
		Miocene	26
		Oligocene	37
		Eocene	57
		Paleocene	65
Mesozoic or Secondary	Cretaceous		141
	Jurassic		195
	Triassic		235
Paleozoic or Primary	Permian		280
	Carboniferous		345
	Devonian		395
	Silurian		435
	Ordovician		500
	Cambrian		570
Precambrian			

PLATE 1 The scale of geologic time, in millions of years before the present, as determined by fossil evidence. Since the development of radioactive dating it has become clear that the period of classical geology, that in which fossils are found, amounts to only a small portion of geologic time.

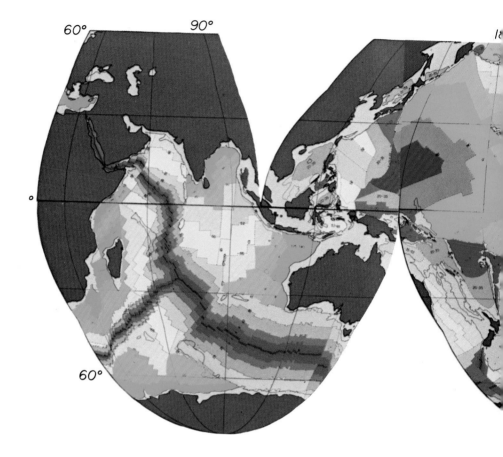

PLATE 2 This is the most modern map of the ocean floor. Each color represents a time interval, as follows:

Dark red: 0–1 M.Y.	Yellow-green: 65–80 M.Y.
Medium red: 1–4 M.Y.	Light green: 80–95 M.Y.
Light red: 4–9 M.Y.	Medium green: 95–110 M.Y.
Orange: 9–20 M.Y.	Dark green: 110–125 M.Y.
Tan: 20–35 M.Y.	Blue-green: 125–140 M.Y.
Light tan: 35–52 M.Y.	Blue: 140–160 M.Y.
Yellow: 52–65 M.Y.	Purple: 160–180 M.Y.

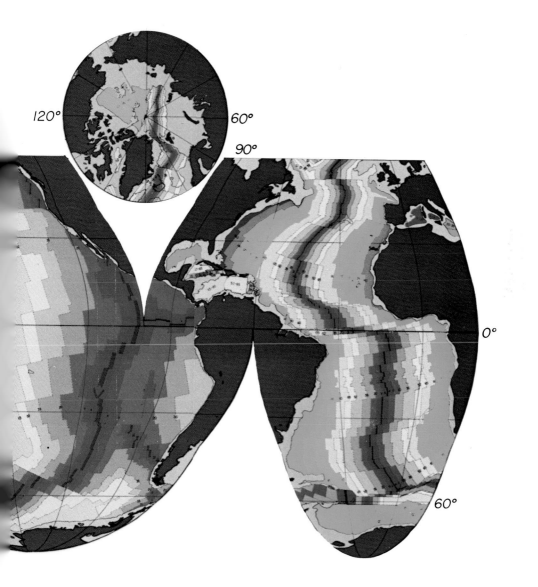

120° 60°

90°

0°

60°

Note that for the so-called rapid ridges, those whose spreading rate is fast, the red color of recent periods is wide (for example, the East Pacific Rise). For slow ridges the red area is narrower (as in the Mid-Atlantic Ridge). The Indian Ocean is particularly interesting. It contains a slow-moving north-south ridge (the Carlsberg) with a narrow red band that meets two other ridges in a triple junction, a slow ridge toward the west and a fast one toward the east. (Map compiled by John Sclater of Scotland and Claude Jaupart of France while they were working together at MIT.)

PLATE 3   This most famous document of the earth sciences is a map of the ocean basins as compiled by Bruce Heezen and Marie Tharp (it includes continental relief as well). Note the continuous network of oceanic ridges, each one deeply crosshatched by numerous transform faults. The rapid ridges (in the Pacific) are much wider than the slow ridges (in the Atlantic). Note also the distribution of the great oceanic trenches (along the coast of Peru and Chile, near the Tonga and Kermadec Islands in the western Pacific, on the shores of Sumatra and Java). The

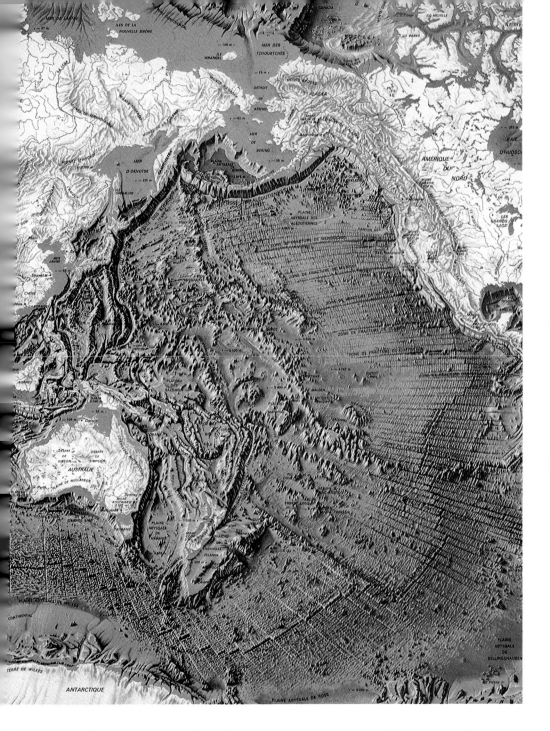

Atlantic has only a few trenches, which are restricted to the arc of the Lesser Antilles and the Sandwich Islands to the south of South America.

Other features are the great abyssal planes and the myriad of volcanoes that populate the ocean floor. Some of these volcanoes are aligned to form chains. The continental mountain reliefs are colored brown; among them can be distinguished the mountain ranges of the Pacific rim, the Alps, and the Himalayan ranges, which are known to have resulted from the collision of the great continental landmasses.

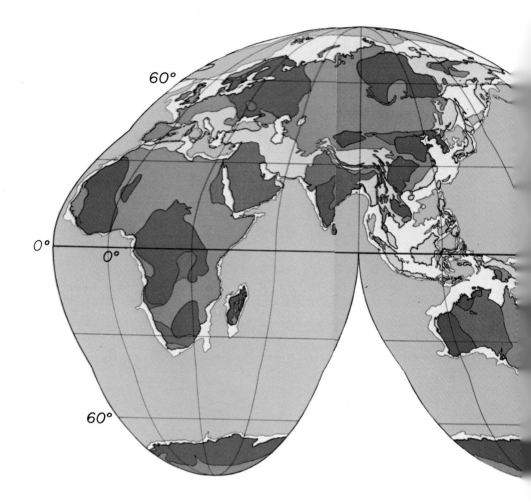

PLATE 4 This map was also created by John Sclater and Claude Jaupart. It shows the continents divided into provinces based on age, using the technique developed by Patrick Hurley. Yellow areas are 0–250 M.Y. of age; orange, 250–800 M.Y.; dark orange, 800–1,700 M.Y.; dark red, more than 1,700 M.Y. of age. Each age province also contains within it older relics. Note that the distribution of the provinces seems to indicate a centrifugal growth of the continents.

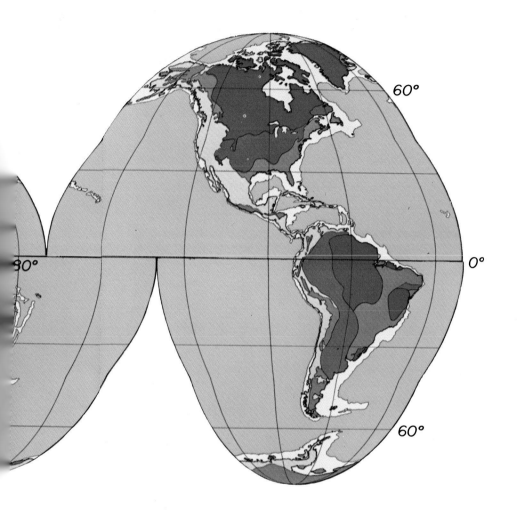

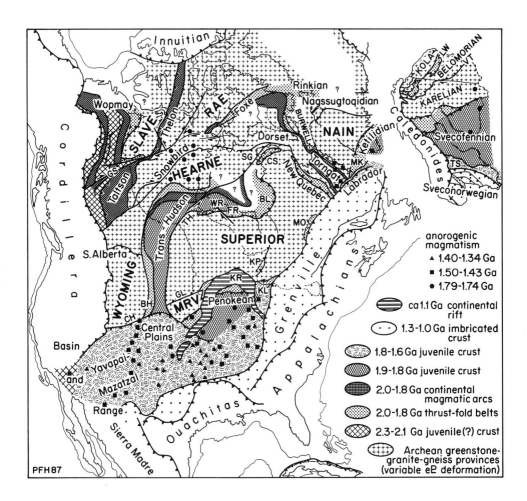

PLATE 5 Geological provinces of North America, mapped according to the age of metamorphic and plutonic rocks. The first such map was made by Al and Celeste Engel; the updated form shown here—which maps the Precambrian tectonic elements of Laurentia (that is, the Canadian shield and its surroundings)—is the work of Paul Hoffman of the Geological Survey of Canada, Ottawa, Ontario. The Baltic shield is shown in a pre-Iapetus reconstruction and Greenland is restored to its position prior to rifting from North America. Upper-case names are Archean provinces; lower-case names are Proterozoic and Phanerozoic orogens. BL = Belcher belt; BH = Black Hills inlier; CH = Cheyenne belt; CS = Cape Smith belt; FR = Fox River belt; GL = Great Lakes tectonic zone; GS = Great Slave lake shear zone; KL = Killarney magmatic zone; KP = Kapuskasing uplift; KR = Keweenawan rift zone; LW = Lapland-White Sea tectonic zone; MK = Makkovik orogen; MO = Mistassini-Otish basins; MRV = Minnesota River Valley terrain; SG = Sugluk terrain; TH = Thompson belt; TS = Transscandinavian magmatic zone; VT = Vetrenny tectonic zone; WR = Winisk River shear zone. The ages of formations are given in *billion years* (abbreviated Ga).

directly. The first study of this type was project FAMOUS (the Franco-American Mid-Ocean Undersea Study). Two groups, American and French, attempted to explore part of the North Atlantic Ridge with a bathyscaphe and two diving saucers. The project was extremely costly and, in spite of what some considered a lot of publicity, the scientific results were rather disappointing at first glance. The exploration showed that, as in Iceland, the ridge had a central valley, 10 kilometers wide, surrounded by steep faults. The floor of the central area consisted of basaltic flows crystallized into pillow lavas. Here and there were conical volcanoes completely analogous to those found in Iceland. Therefore, the Icelandic model seemed to be confirmed. Many researchers would have stopped there, but Jean Francheteau of France persevered. Simplifying the technology and seeking more American partners, he assembled a group with Bob Ballard of Woods Hole and Harmon Craig of the Scripps Institution; they explored not simply a piece of a ridge, but a number of ridges.

The second expedition of this type, whose code name was CYAMEX (after Cyana-Mexique), took place off the coast of Mexico. The area to be explored was the East Pacific Rise near the mouth of the Gulf of California. This part of the East Pacific Rise is spreading much more rapidly than the FAMOUS zone: the spreading rate in the Atlantic is 1.5 centimeters per year; in the zone studied, called RITA (after the Rivera-Tamayo fault zone), the rate is 6 centimeters per year. In this part of the East Pacific Rise there is no longer a central valley. The ten kilometers on each side of the central magnetic anomaly presently form a swelling or bulge. The volcanic structures are much more complex, and the distinction between cones and lava flows is blurred—or, rather, the two types of structure are much more closely mingled. Moreover, there are real submarine lakes of lava, indicating highly productive volcanic episodes.

But the most remarkable finding of this expedition was certainly the discovery of metallic sulfides laid down in the seawater-basalt interface. The existence of veins of metal, of actual "mines" linked to pillow lavas, has been known since antiquity. The island of Cyprus is famous for its copper mines (the name of the metal is derived from that of the island). The veins consist of accumulations of sulphur associated with a gigantic ophiolitic massif, the Troodos Massif, the sulphurs being closely associated with the lavas that flowed out into basaltic pillows. Thus what had been hoped for, on the basis of observations of ancient terrains (180 million years old), has been found in present-day conditions.

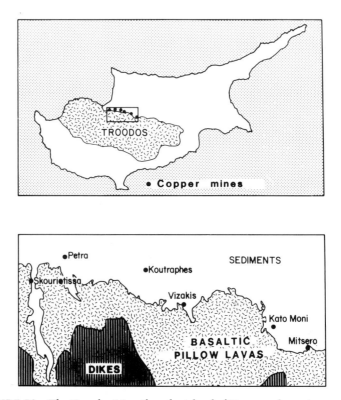

FIGURE 58   The Troodos Massif on the island of Cyprus, *above,* is composed
partly of ophiolites. The area of detail, *below,* shows the distribution of copper
mines. The copper mines, control of which is a source of friction in the po-
litical conflicts in Cyprus, are located at the top of the ophiolite section, particularly at the contact between basaltic pillow lavas and overlying sediments.

The systematic exploration of this phenomenon led the American and French groups to discover a much more general process of
which the deposit of sulphur is only one manifestation: the
circulation of hot water through the oceanic crust in ridge areas.
Seawater penetrates newly formed basalts, infiltrates them, dissolves certain elements, and eventually attains a temperature of
300° Centigrade, at which (under the pressure found on the ocean
floor) it vaporizes. Jets of vapor containing chemical elements
taken from the oceanic crust rise toward the surface and are
injected into the ocean. The fact that elements from the mantle are
injected into the ocean is very important to the understanding of
the chemistry of seawater. It was generally thought that the
chemical composition of seawater was the result of two comple-

mentary phenomena: the contribution of river water (that is, the erosion of continents) and sedimentation (the extraction of insoluble salts). Now it was necessary to add a third element: the mantle. John Edmond of MIT attempted to quantify this phenomenon. According to him the hydrothermal circulation at the ridges is so intense that the entire ocean is recycled through the oceanic crust in 10 million years. It is therefore a phenomenon of considerable importance.

For the sake of historical accuracy, it must be mentioned that the circulation of hot water at the ridges was not unexpected. A theoretical conjecture had been made about the deposits of copper on Cyprus. When Jan Bottinga and I calculated the continuous dynamic model of a ridge, we realized that the flux of heat engendered in such a model was two or three times greater than that actually measured at the ridges. Following a suggestion by John Elder of New Zealand, we postulated the existence of a seawater circulation system that carried away part of the heat transferred by the magmas. At the University of Washington Cliff Lister had made a similar hypothesis. Nevertheless, it was necessary to await the detailed study of the Galápagos Ridge by the Scripps Institution in 1977 to observe directly the existence of submarine geysers. The following year the CYAMEX expedition discovered the submarine sulphurs.

Exploration of the ridges did not stop at the RITA zone. Franco-American groups continued their effort to study even fast-spreading ridges. The harvest of results continued with exploration of the Nazca zone, off the coast of Peru. Hydrothermal circulation was found here and there, confirming the specific morphology proposed for fast-spreading ridges.

At this point we have arrived at the limit of present-day research, and a little time must pass before the results can be appreciated. Let us turn now to what lessons can be drawn from the geology of ridges.

The Geology of Ridges

The geology of ridges opened a new chapter in the earth sciences. Scientists are still writing its introductions. I shall try, however, to pick some of the most important results out of the plethora of high-quality studies. In my opinion, fundamental progress has radically transformed our picture of the ridges.

For a long time our object was to model the processes that occurred at ridges, but the ridge was considered a universal type

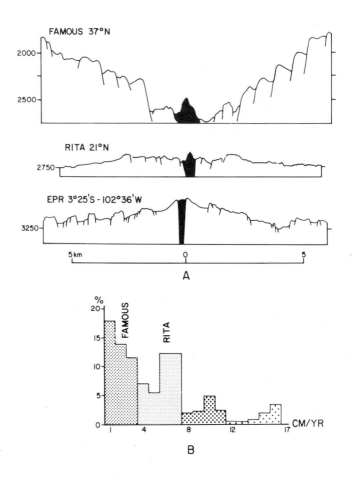

FIGURE 59 (A) Topographic/morphologic profiles of spreading centers that open at different spreading rates. In each profile, the top line shows the depth to the ocean floor, lines extending beneath the topographic outline show the estimated position of normal faults, and solid black shows the extent of the active axial zone of volcanism. Shown here are profiles of the FAMOUS region of the northern Mid-Atlantic Ridge, where the opening rate is 2 centimeters per year; the RITA region of the East Pacific Rise in the Gulf of California, where the opening rate is 6 centimeters per year; and the southern East Pacific Rise, which is spreading at 10 centimeters per year. Note that at slow-spreading ridges a large central valley is present and that small, central volcanic highs occur at faster-spreading ridges.

(B) Percents of ocean ridges spreading at a given speed. Mid-oceanic spreading rates range from 0.5 centimeter per year in the Arctic Ocean to 17 centimeters per year in the southern East Pacific Rise—that is an increase of a factor of 35. (Figure after J. Francheteau.)

of structure: we believed that we should find a unique paradigm. At present we are interested in various types of actual ridges, no longer just in theoretical constructs. Each ridge has a specific spreading rate, a particular geometry, and its own history. Its structure and functioning are specific to it. Now we can begin to compare ridges, link their variations to physical parameters, and start to understand the mechanisms at work in them.

It also became clear that a discontinuous, episodic model must be substituted for the plate-tectonic model of continuous creation of the lithosphere. The spreading of the ocean floor appears continuous on a time scale of millions of years. On the scale of one year it is a succession of rapid phenomena separated by periods of quiescence. This is also the case for earthquakes and lava flows. In Iceland this episodic volcanism and seafloor spreading, which is being measured today, can be seen in the geologic record. It is only on the global scale and on the geologic time scale that spreading appears to be continuous, with an *average* velocity of a few centimeters per year. The accumulation of small discontinuities creates the illusion of continuity. The differences in spreading rates measured on various ridges are not a product of differences in the rate of the rising of materials from the mantle, but much more of the existence of longer or shorter periods of quiescence separating the active phases. These findings are as important for the chemical phenomena that take place in the ridges as for the occurrence of earthquakes or the birth of volcanoes or transform faults.

The Geology of Transform Faults

Transform faults are the only plate boundaries that conserve the surface of plates. They do not create surface as ridges do, nor do they destroy it as trenches do. Their orientation is directly linked to the position of the pole of rotation between adjacent plates, and they make it possible to locate such a pole, as Morgan noticed at the birth of the plate-tectonics theory. So it is not astonishing that those who study geodynamics devoted a large part of their efforts to transform faults. At sea many cruises were devoted to the study of the great transform faults, such as the Gibbs Fault or the Vema Fault in the Atlantic and the Clipperton Fault or Sequieros Fault in the Pacific. Submersible diving programs, notably project FAMOUS, also focused on faults. Unfortunately, it must be said that nothing spectacular or very clear came out of these studies. It would have been nice to see zones of fissures affirming the existence of a special sort of volcanism along faults; but it seems that if this phenomenon exists, it is very discreet and not linked to spectacular volcanism. Certainly faults show considerable vertical offsets, throws that combine, as is always the case with faults, with the horizontal shifts already mentioned, and these escarpments allow us to detect deep cracks in the oceanic crust. So they are very useful structures. Of course, the mechanism of their formation at the ridges has been the object of interesting theoretical models, but no all-embracing interpretation has developed around them. Perhaps this is still to come.

Because it is easier to observe faults on the continents than at a depth of 3,000 meters, the detailed study of transform faults has been much more active on the surface—especially since the San Andreas Fault, which crosses California from south to north and which had been the classic fault described in all the American geology textbooks for decades, was classed among the transform faults when plate tectonics was in its infancy. Geologists started looking for transform faults everywhere; all the great strike-slip faults in the world were rapidly placed in that category: from the Jordan Fault in Palestine to the Altyn-Tagh in Asia, including the Alpine Fault in New Zealand, the Chaman Fault in Pakistan, and the North Pyrenean in France. A real "transforming" mania possessed the structural geologists, and it was good form during this period to call this or that fault a transform fault.

But all the great faults are not transform faults, only those located on active plate boundaries are. Many are on borders that have been reactivated, their previous state not having been that of faults, but more often of ocean-continent borders. Although, for example, the North Pyrenean Fault or the Chaman Fault are transform faults—the first joining the spreading zone of the Gulf of Gascony to a subduction zone under Italy, the second joining the Himalayan subduction zone to the Indian Ridge—others that were classed as transform faults, like the Altyn-Tagh Fault, were not active plate boundaries at all.

Active research on transform faults did not bring to light any new characteristics, any more than research on the ocean floor had. Nevertheless, it reminded us of the importance of horizontal displacements for tectonics, and that is not a small benefit. Faults had long been neglected, poorly mapped and poorly observed; the noble geologic structure was thought to be the fold—more generally, the great folded mountain ranges such as the Alps. Faults seemed to lack any great significance. This error had been pointed out and fought

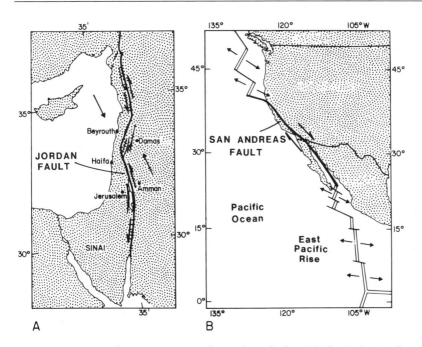

FIGURE 60   Two famous continental transform faults. (A) The Jordan Fault joins the Red Sea spreading center to the collision zone in Iran. (B) The San Andreas Fault joins the East Pacific Rise in the Gulf of California to the Juan de Fuca Ridge in the north.

against by Argand as well as Wegener, but it had remained entrenched in the minds of most field geologists. Their refusal to acknowledge the importance of the horizontal strike-slip fault was in fact a rejection of the idea of continental mobility.

Subduction Zones

The study of subduction zones did not begin with plate tectonics, as had been the case with ridges and transform faults. Subduction zones had always been of interest to geologists and geophysicists, because earthquakes and spectacular volcanic eruptions originate in them. Of course, there was no talk of "subduction"—the regions were called "mobile zones" or "active belts"—but everyone in the earth sciences knew that phenomena important for terrestrial dynamics took place there. This general curiosity was amplified by a particular sociological fact. Japan, a great scientific country, is situated next to a subduction zone. For a hundred years the activity of Japanese researchers has been directed toward the study of the subduction process.

In the plate-tectonic theory, subduction zones are the complements of ridges. They are areas in which plates disappear and are swallowed up by the mantle. The mode of operation of these zones had been elucidated in 1930 by Felix Vening-Meinesz of Holland, who had developed a method for measuring anomalies in the field of gravity on the seafloor. The technique was difficult: changes in the gravitational field are very slight, so the measurements had to be very precise, and the marine environment is not particularly favorable for exact measurements. Nevertheless, Vening-Meinesz managed to overcome these difficulties by taking his measurements from a submarine. Working near Indonesia, he established that the great oceanic trenches are characterized by gravitational anomalies, which he interpreted by postulating the existence of convection currents in the mantle, the trench zones being the location of descending currents. This model is not far from that of the "plunging plate."

Meanwhile, Japanese geologists were preoccupied by the relationship between earthquakes and volcanic eruptions or, more generally, between seismicity and volcanism. A. Sugimura, a young Japanese volcanologist, and then Hisashi Kuno, the master of volcanic petrology, suggested that the origin of lavas is the Benioff-Wadati zone, the zone in which deep earthquakes are localized. I call this Kuno's conjecture.

Arthur Holmes took up all these ideas in his *Principles of Physical Geology* (1945) and constructed a synthetic model of a

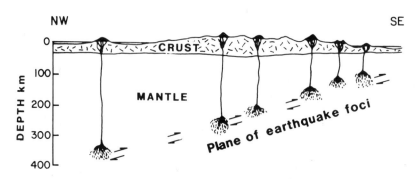

FIGURE 61 Cross-section of Japan from the southeast to the northwest, illustrating Hisashi Kuno's idea that the region of subduction-zone earthquakes is also the region of magma generation that causes surface volcanism.

subduction zone that has hardly been superseded today, even if people often pretend to have forgotten it.

Earthquakes as Tracers of Plunging Plates

Subduction zones are seismologically unique in that they are the site of deep earthquakes, those whose depth is greater than 150 kilometers. The oceanic plate is cold, because its surface has spent hundreds of millions of years next to seawater whose temperature is 4° Centigrade. The mantle, on the other hand, is hot. The thrusting of a cold plate into the hot mantle stimulates two types of reactions: a thermal phenomenon, the reheating of the plate with associated "cracking," and a mechanical phenomenon, the resistance of the mantle to penetration by the lithospheric plate. These two phenomena work together to cause breaking in the interior of the cold, rigid, and brittle lithospheric plate. Calculations show that even after several dozen million years the plunging lithospheric plate remains much colder than the mantle that surrounds it. The surrounding mantle is more plastic; it reacts to mechanical constraints not by breaking but by slow deformation. The different reactions of the plate and the mantle explain why only plunging plates or subduction zones are the sites of deep earthquakes.

Using these findings in reverse, seismologists can locate and study the properties of plunging plates. It is immediately apparent that the subduction zones are not uniformly distributed in relation to the great physiographic borders of the earth. Sometimes subduction takes place at the ocean-continent border, as in Mexico, Chile, or Peru; sometimes, on the other hand, it takes place in mid-ocean, as is the case near the Marianas, Tonga, and Kermadec islands or in the Lesser Antilles. Finally, subduction sometimes takes place along the edge of a continent but in a more complex arrangement, such that between the subduction zone and the continent there is a rather well-developed ocean basin. There may also be an archipelago of islands or microcontinents in this basin. This arrangement, called an *island arc*, exists along almost the entire western circumference of the Pacific, from the Aleutians, the Kuriles, and Japan to Indonesia, New Guinea, and New Zealand.

The Lamont geologists active in this area of research (Bryan Isacks and Peter Molnar in particular) attempted to explain the subduction zones by studying the earthquakes that occur there. They ascertained that the geometric characteristics of the zones in which earthquakes, and particularly the deep earthquakes, are

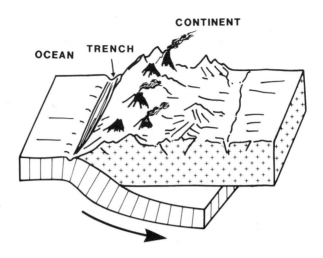

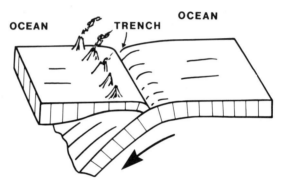

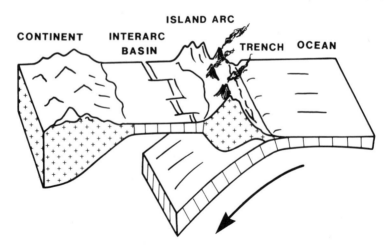

FIGURE 62   The three fundamental plate geometries of subduction zones. Note the presence of volcanism in all cases.

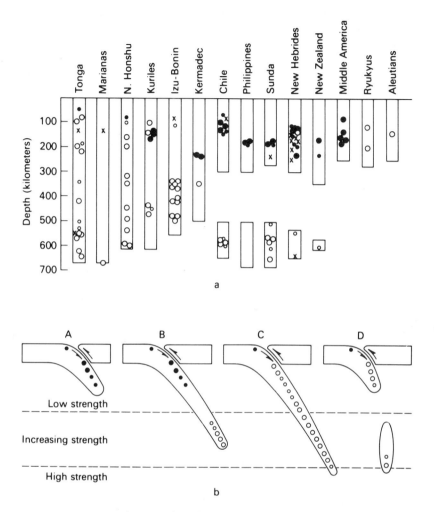

FIGURE 63 Distribution of earthquakes as a function of depth. The top section shows the distribution of different types of earthquake along the Benioff-Wadati zones associated with subduction zones around the world; there may be gaps in the zones. Filled circles indicate tensional earthquakes, open circles indicate compression. The lower section is a schematic diagram of Isacks and Molnar's explanation for the observed distribution of earthquakes.

produced vary from one region to another. As I noted before, this zone, called the Benioff-Wadati zone in seismology, is in fact the zone where the cold plate plunges into the mantle. Earthquakes attain their maximum depth of 600–700 kilometers in several regions of the earth (Tonga, the Marianas, Japan, the Kuriles).

Elsewhere the maximum depth is less—300 kilometers in Central America, 450 kilometers in New Zealand, 250 kilometers in the Aleutian Arc. The angle that the Benioff-Wadati zone makes with the horizontal varies also; sometimes it is slight, as in the Aleutians, and sometimes very steep, as near the Tonga Islands or the Marianas Arc.

Isacks and Molnar used detailed studies of the foci of earthquakes to show that when the plate is beginning its downward plunge, earthquakes are a result of the bending of the plate; when the plate has reached a depth of 500 kilometers, earthquakes result from resistance to penetration—that is, they are the compressive type of quake. These discoveries were put to use in understanding the mechanism of subduction itself.

Volcanism

Along with earthquakes, volcanic eruptions and, therefore, volcanism are features of subduction zones. Volcanism in subduction zones is very different from that on oceanic ridges. It is more violent, characterized by explosive emissions or avalanches of cinders that are called nuées ardentes (French for "glowing clouds"; in English a comparable term is *ash flows*), such as those that ravaged Saint Pierre on Martinique in 1902, killing many people. The most destructive eruptions in man's memory were situated in subduction zones, including the famous Krakatoa eruption, which swallowed up an Indonesian island in 1887, and the Santorini, which no doubt destroyed the civilization of Crete in 1400 B.C.

Most often subduction-zone volcanoes have a spectacular cone-like shape and a regular distribution; for a given zone, they have periodic spacing of 30 to 60 kilometers. For the great majority the nature of the products emitted—bombs, lavas, cinders—is very different from that of ridge-zone volcanoes. The debris is much richer in silica and therefore more like the materials that characterize the continental crust. The most common rock and the one that dominates the emitted products in terms of volume is *andesite*. Its chemical composition is similar to that of the average continental crust.

From this duality in the nature of volcanic products it is only one step (and many take it without hesitation) to stating that volcanism at ridges produces basalts and creates the *oceanic* crust and that volcanism in subduction zones produces andesites and therefore creates the *continental* crust.

This theory, which makes volcanism the creator of the earth's crust, is very appealing. Two types of volcanism, that of the spreading zones or ridges and that of the subduction zones, correspond to two types of materials, two types of crust—the oceanic and the continental. How could such simplicity not be attractive? But it still does not explain why there is volcanism in subduction zones.

At the ridges a piece of mantle becomes decompressed and has a tendency to melt as it rises toward the surface. But in subduction zones, a piece of cold mantle is being buried and therefore compressed. These factors, both the temperature and the change in pressure, seem to contraindicate melting, but still there is volcanism!

It must be remembered that a "sandwich" has been created at the ridge crest. A layer of basalt was rapidly formed, hydrated, and "wedged" between marine sediments above it and a piece of residual mantle below it. The melting temperature of basalt and of the sediments is much lower than that of mantle rock. In subduction zones, however, the sandwich (basalt plus sediment) plunges into the mantle and therefore is reheated (slowly but inexorably). When it reaches the melting temperature of the mixture (basalt plus sediment plus water), the sandwich melts and gives birth to volcanic magma. The magmatic liquid, being less dense and therefore lighter than the surrounding mantle, tends to rise to the surface and erupt as a volcano.

This mechanism explains why the zone in which volcanoes appear is always situated some distance away from the trench, where the plate begins its descent. A certain time and therefore a certain distance (covered at a speed of a few centimeters per year) is necessary for the plate and its sediments to be heated to the melting temperature. The oceanic crust is impregnated with water during its underwater journey, and the sediments themselves are full of water. The magmas produced by this melting are therefore rich in water, which as it is heated is transformed into vapor. Given these conditions, it is easy to understand the explosive power of andesitic volcanoes. Kuno's conjecture has been given a logical and coherent physical explanation.

Andesite is not the only volcanic rock formed in subduction zones. There are also smaller proportions of basalts and melted granites, called rhyolites. The simple idea that basalt is formed when the plate alone melts, that rhyolite is formed when sediments melt, and that andesite is formed when a mixture of basalts and sediments is melted arises from this symmetry, but it is far from unanimously accepted; indeed, it is the subject of passionate debates among petrologists.

If we compare the volcanic characteristics of several subduction zones, as we did for earthquake zones, we can see that the zones vary widely. The proportion of the types of rocks, the distances from the volcanic front, and the frequencies of eruption are all variable. Therefore, melt is produced in forms that differ according to local conditions. Among the determining conditions is the type of constraint involved, and whether the subduction zones are in compression or in extension.

The Tectonic Nature of Subduction Zones

As I have pointed out before, subduction zones appear in the morphology of the ocean floor as deep trenches. Some trenches in the Kurile Islands attain depths of 11,000 meters; bathyscaphe diving records were broken there. Now, exactly why do these trenches exist? Early on it was supposed that the plunging of a plate into the mantle would meet with resistance and that the plate, as a result, must be pushed down; the subduction front was therefore assumed to be in compression. This idea seemed consistent with the fact that there are ranges of folded mountains around the circumference of the Pacific, whose folds implied compression.

Furthermore, sedimentary accretion prisms at subduction zones were explained by compression at the zones. If the plate enters the mantle by force, it is unlikely that the soft sediments that it carries would be pulled into the mantle. They would instead be scraped off and form a highly pleated hump. This prism is added to the andesites to give birth to new pieces of continent.

Observations in parts of the western Pacific and especially in the Tongas and the Marianas, however, showed that subduction zones were *not* in compression. Mechanisms at the foci of earthquakes were the first indicators of this, but they were reinforced by sedimentological observations. There were hardly any sediments in the trenches. What had happened to the hundreds of cubic meters of sediments supplied annually by the conveyor belt of the ocean floor? Remember that this enigma had led Harry Hess to the hypothesis of seafloor spreading.

The conclusion must be that the sediments *are* swallowed up in the mantle with the floor. But how could these soft sediments be swallowed up if the area is not in extension? Studies seemed to be inconclusive on this point. To reconcile the conflicting observa-

Seiya Uyeda

tions, Seiya Uyeda of Japan proposed that there must be an evolution in the types of subduction.

Subduction begins, as in Chile today, with a break-in by force; the angle of the Benioff-Wadati zone is slight, and the trench is bordered by folded sediments. As the plate is gradually buried, it bends and, somewhat like a person swimming the crawl, "pulls" backward in the mantle underneath the plate. Little by little the trench becomes an extension zone. In summary, the subduction zone collapses into the mantle carrying the sediments on its back into the depths with it. Over the course of time this collapse increases in scope, and therefore the subduction zone marked by the oceanic trench has a tendency to recede, accentuating the extensive character of the zone. Such would be the case of the Marianas today. Uyeda classified all the known subduction zones between these two extremes. One can see the potential richness of the concept of subduction in this model. Ridges are areas of unidirectional transfer from the depths to the surface. Subduction

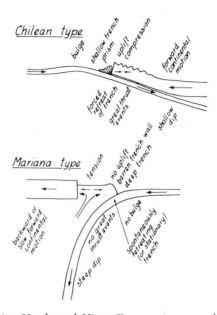

FIGURE 64 Seiya Uyeda and Hiroo Kanamori proposed a model relating subduction, plate motions, and the resulting evolution of subduction zones. In Chilean-type subduction both the continent and oceanic plate converge toward the trench. The result is a shallowly dipping Benioff-Wadati zone and compressional mountain building within the continent that is overriding the oceanic plate. In Mariana-type subduction the overriding plate is retreating from the subduction zone. This leads to a deeply dipping Benioff-Wadati zone and extensional spreading behind the trench.

zones are the site of bidirectional exchanges: from the surface to the depths for the plate and its sediments, from the depths to the surface for the magmas. Moreover, subduction zones exhibit lateral variation according to their stage of evolution.

Heat Flux and Metamorphic Belts

The distribution of the heat flux in subduction zones follows a predictable pattern. At the oceanic trench the flux is very weak; it increases abruptly near the zone of active volcanism. In this area the flux of heat is very high: more than fifteen times the terrestrial average value. Little by little it becomes normal again as one moves away from the trench and the volcanism dies out. This distribution is easy to understand if one remembers that the trench corresponds to a transfer of the cold plate to the interior and that the volcanic zone corresponds to the transfer of hot magma to the surface.

The first geologic use for this observation was developed by A. Miyashiro of Japan and Gary Ernst of the United States to explain metamorphism in mountain ranges. Metamorphism, in the geologic sense of the term, is the phenomenon by which a given rock, whether sedimentary or plutonic, is transformed into a new rock by crystallization or recrystallization. The primary rock is crystallized into a metamorphic rock. A feldspar sandstone is transformed into gneiss, a limestone into marble, a clay into schist. Metamorphosis takes place because of the changes in temperature and pressure to which the rocks are subjected. The nature of the minerals that are created reveals the conditions of temperature and pressure that caused the transformations, thanks to the mineralogic code of metamorphosis mentioned in Chapter 2.

Miyashiro and Ernst noticed that two belts of metamorphic rocks can be defined in Japan and in California. The belt nearer the ocean is characterized by a metamorphism that develops under high pressure but relatively low temperature (300° Centigrade). The interior belt is characterized by high temperature (300–600° Centigrade) and medium pressure. Ernst and Miyashiro attributed the "cold" metamorphism to the sinking of the sediments in the oceanic trench. With the accumulation of sediments, the pressure increases but the temperature remains relatively low (because the sediments are in contact with the cold, plunging plate). The internal metamorphic belt is linked to reheating associated with the intrusion of basaltic and andesitic magmas. These multiple heat sources cause the adjacent continental crust to heat up.

Thus the study of the metamorphic belts enables us to reconstruct ancient thermal regimes, which have been "imprinted" in the various metamorphic minerals, and to recognize the existence of ancient subduction zones when seismicity has completely disappeared. Paired metamorphic belts and andesitic volcanic rocks, in other words, are plate-tectonic indicators from which geologists can reconstruct the past.

Marginal Basins

Around 1975 the center of interest suddenly switched to a structure that had not at first greatly interested the modern advocates of mobility theory but that had attracted Wegener's attention: basins situated behind the island arcs, called marginal basins. The plunging plate had been the center of attention; henceforth the little wedge above it would become important. Two scientists, Dan Karig, who was then a young student at Scripps, and Seiya Uyeda, pioneered the study of the island arcs.

Karig discovered that in the basins behind the arcs of Tonga there is a basaltic floor similar to the floors of the great oceans. Moreover, he showed that the tectonics and the sedimentation of the basins suggest an extension structure accompanied by highly developed volcanism. He hypothesized that these marginal basins are born as a result of the activity of secondary oceanic ridges situated parallel to the trench but behind the arc. Arthur Holmes had already imagined such an arrangement. Uyeda quickly showed that this arrangement is not unique to the Lau basin near Tonga. For almost every subduction zone in the Western Pacific he discovered magnetic anomalies resembling Vine-Matthews-Morley anomalies in the basin behind the Marianas Arc. Moreover, he showed that some of these ridges are young and that young ridges are marked by anomalies in the heat flux. In older ridges volcanism has been extinguished and heat flux is weak. It is therefore more difficult to detect older ridges geophysically.

From these developments it can be concluded that adjacent to the subduction zones, where the lithospheric plates plunge into the mantle, there are structures similar to the oceanic ridges where additional plate creation takes place along with volcanism and heat transfer. This volcanism is very different from that of the Benioff zone and very similar to that of a typical oceanic ridge. Therefore it is a volcanism whose source is not the plunging plate, but the mantle wedge above it.

Next, mechanisms to explain this phenomenon were conceived: physical mechanisms involving the existence of convective cells were proposed; and chemical mechanisms were postulated on the notion that the plunging plate loses water and that this water migrates vertically and promotes melting above the subducting plate. But at this stage the precise mechanism is not important. What does matter are the basic observations and the logical connections that unite them. The existence of structures that create surface (ridges) connected to structures that destroy surface (subduction zones) has a fundamental logical consequence: Over the long run many subduction zones are going to absorb completely the plates that carry them. If all subduction zones act in this manner the number of plates on the surface of the earth will decrease continuously. Eventually there will be only one plate on the earth's surface. Actually the subduction zone "corrects" this contrary tendency by creating a *secondary ridge* behind the island

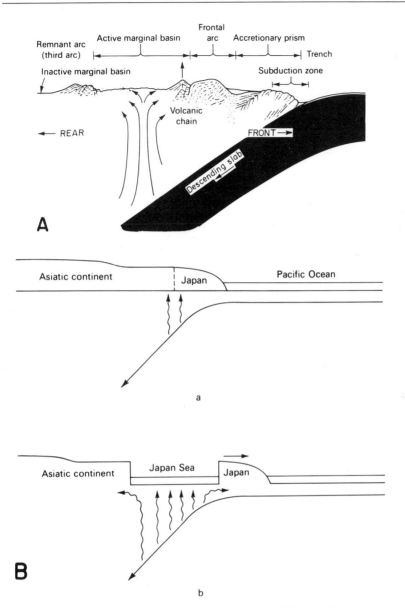

FIGURE 65 The complementary models of marginal basins by Dan Karig and Seiya Uyeda. (A) Karig, a geologist, presents a sketch of the essential surface structures of a marginal basin with deeper structures shown only when they help to explain surface observations. (B) The geophysicist Uyeda, on the other hand, in his model of the formation of the Sea of Japan shows only the essential mechanisms leading to marginal-basin formation. These two figures illustrate the two opposing tendencies of modern geology: deductive (A) vs. inductive (B) geodynamical reasoning.

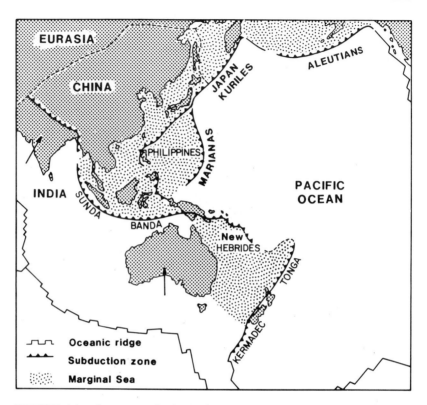

FIGURE 66   The principal island arcs and marginal basins of the western Pacific.

arc. A secondary ridge can eventually become an oceanic ridge and produce two large-sized plates. Are secondary ridges the regulatory phenomena that assure the permanence of plate tectonics?

Another consequence is geologic. The complexity of island arcs—with their volcanism, varied sediments, and continental islands—reminds field geologists of formations that they have seen in mountain ranges. Are island arcs the long-awaited key to the complex mysteries of mountains?

Mᴏᴜɴᴛᴀɪɴs, the most prominent geologic feature of the surface of the earth, have always attracted men. Some want to conquer them by climbing them; others, by understanding their origin. This attraction alone is enough to explain the success of tectonics, the branch of geology dealing with the massive structures of the earth's crust. The mystery of the mountains became even more profound when fossils of marine animals were discovered at heights of 3,000 or 5,000 meters. Forces capable of raising the sea bottom and lifting it to great heights must therefore exist, but what are these forces?

From a more technical point of view, mountain ranges offer the geologist the opportunity to visualize the three-dimensional structure of the earth. A Himalayan valley with 7,000 meters of relief (valley at 1,000 meters above sea level, peaks at 8,000 meters), for example, is in effect a slice cut into the earth, the equivalent of a bore hole drilled to the same depth (and not a single bore hole, but a continuous trench!). Geologists refer to the process of mountain formation as *orogeny*, a term derived from the Greek for "genesis of relief."

A century of study has made use of all available techniques of classical geology, including dating the strata by fossils and mapping the different terrains as well as the folds and faults. This research has yielded several important discoveries: the mountains have been clearly fixed in space and time. The Himalaya were formed in the Cenozoic era (less than 60 ᴍ.ʏ.ʙ.ᴘ.); the Rocky Mountains, during the Mesozoic (210 to 60 ᴍ.ʏ.ʙ.ᴘ.); and the Appalachians and the Urals, in the Paleozoic (between 550 and 200 ᴍ.ʏ.ʙ.ᴘ.).

The structure and composition of mountain ranges have been thoroughly studied as well. Their most spectacular characteristic is the folded and faulted strata that form a high relief of deformed rocks, but the core of a mountain belt consists of metamorphic rock and granites, which probably account for a major part of its volume. The highest summits of the Himalaya, such as Manaslu,

Makalu, K2, and Everest, are composed of granites and gneisses, despite the small sedimentary layer on the top of Everest. The ranges of North America, including the Appalachians, are full of granites, too. This is especially significant because granites are igneous rocks, products of the solidification of hot magmas, whereas metamorphic rocks, such as gneisses, are rocks transformed by heating. Dating of granites and metamorphic rocks has shown that their formation occurs more or less simultaneously with folding and faulting. Mountain building appears to be a multifaceted phenomenon involving relief generation, heat transfer, and rock formation.

The Geologic Cycle and Geosynclines

The search for an explanation of the formation of mountain belts in light of what was known about the origin of the different types of rocks led geologists to build a synthetic view of geology around the mountain-building process. Two important concepts are linked to this view: the geosyncline and the geologic cycle. They have dominated geologic thought for more than fifty years.

The concept of the geosyncline was introduced in 1859 by James Hall and later taken up by James Dwight Dana, American geologists who developed a model to explain the Appalachian range. A geosyncline is a marine trough that has been filled in with sediments. After the trough is filled in, its basement rock sinks under the weight of the sedimentary load (as is the case for all basins that accumulate sediments). This is called *subsidence*. The sediments, which have been progressively transformed into hard rock as water was squeezed out of them, are raised toward the surface. Their rise is accompanied by a horizontal shortening of the trough that forces the sedimentary rock to fold over upon itself, similar to what happens when the water is wrung from a wet towel. The folding is accompanied by thermal phenomena and by the intrusion of magmas from the earth's interior that lead to the creation of a hard and rigid piece of continent. Once the tectonic phenomena have ceased, an isostatic readjustment that reestablishes gravitational equilibrium raises the structure built in the trough to a great height. The basic idea, therefore, is that the mountain range is a former sedimentary basin that has been transformed into an elevated structure.

Once the mountains have risen, their newly formed peaks become subject to erosion. Debris and alteration products, torn

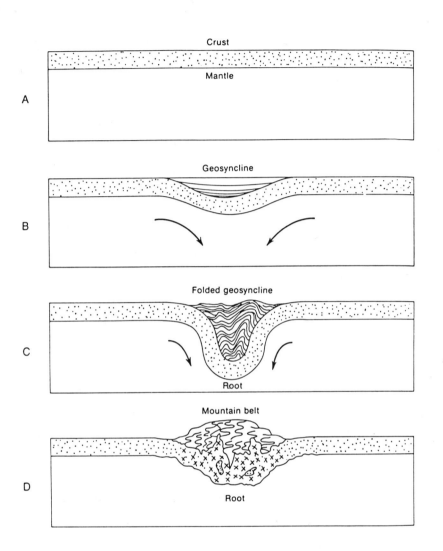

FIGURE 67 David Griggs's schematic picture of geosynclinal evolution and mountain building. In this prescient model Griggs postulated that mantle convection could lead to formation of a basin that would first fill with sediment and eventually be subducted, folded, and compressed by underlying convection into a mountain belt. Note, however, that except for the driving force, the cooling-and-contracting-earth theory would predict essentially the same pictorial evolution of a geosyncline.

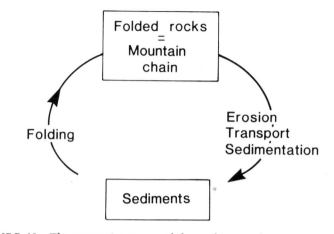

FIGURE 68   The successive stages of the geologic cycle.

from the summits by storms and carried along by rivers, sediment
out and accumulate in new troughs, new geosynclines, that will in
turn be folded, intruded by granites and metamorphosed, as the
cycle repeats itself indefinitely. The geologic life of the planet
appears to be divided into two episodes, as distinct from each other
as they are complementary. One is of long duration, slow and
inexorable: the erosion of the peaks and transport of their products
into the sea. This is the sedimentary episode, the filling of the
oceanic basins. The other is active, violent and short-lived: sediments are folded, heated, hardened into granites, raised into peaks,
and finally turned into pieces of continent, mountains from the sea
bottom. Of course, the second episode, orogeny, is the more
interesting. Therefore many geologists thought that the way to
solve the essential mystery of geology was to study the structure of
mountains and the way in which they are folded. For them
tectonics is at the heart of the geologic debate; it is the key
discipline. The theory of geosynclines achieved a "complete"
synthesis of geologic phenomena, from mountain building to
igneous rock formation and the processes of erosion and sedimentation. The forces that turned a trough into a peak, however,
remained mysterious!

Mountain-Building Forces

Hall and Dana believed that the contraction of the edges of the
geosyncline was a consequence of the cooling of the earth. This

seemed to fit in neatly with the baked-apple theory. At the beginning of the twentieth century, however, geologists expressed the greatest doubts about this theory even though the calculations of Sir Harold Jeffreys (still the same guy!) seemed to "prove" it.

Between 1884 and 1889 the French geologist Marcel Bertrand discovered the great lateral displacements of the folded rocks he called thrust sheets. His work increased skepticism of the baked-apple theory. Franz Kossmat, Émile Argand, and Pierre Termier showed that the Alps are a range of gigantic complex folds and thrust sheets and stressed that such a structure necessitated shortenings on so great a scale that they would be difficult to explain by the cooling-off theory. As we know, Wegener had used the same arguments. Argand drew the "logical" consequences of this argument and bravely adopted Wegener's theory. Few would follow his lead.

Later on, when Wegener's theory of continental drift was abandoned, this explanation of mountain building disappeared also. The concept of the geosyncline remained, however, and geologists

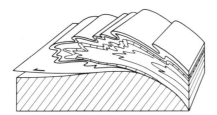

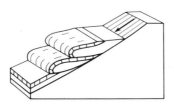

FIGURE 69   Two rival explanations for the existence of large lateral displacements of folded terrains called thrust sheets. In the "Alpine" interpretation, *top,* a thrust sheet of folded rocks is squeezed out of the root zone over neighboring rocks. The lateral movement is a consequence of the extreme folding and shortening. In the gravity-sliding interpretation, *below,* the folding occurs during lateral movement driven by gravitational forces, causing the thrust sheet or nappe to slide over the substratum.

all over the world continued to study mountain ranges, to map them, to measure the length of their folds, and to verify the existence of extensive shortenings without wondering about their cause. Mountains, their structure, and their form became an object of study in themselves. The cause of the folding no longer seemed to interest geologists.

In their synthetic studies of mountain ranges and geosynclines Jean Aubouin of France, T. Matsumoto of Japan, and Marshall Kay of the United States attempted to understand the logic in the progression of episodes, to reconstruct successive geographies, and to compare various mountain ranges, but they deliberately ignored the problem of what caused the folds. These attitudes are representative of the geologic community from the 1950s to 1970s, but a few geologists and geophysicists escaped this retreat into taxonomy.

For example, in 1939 David Griggs at the University of California at Los Angeles offered a dynamic explanation of geosynclines based on convection currents and carried out laboratory experiments to illustrate his theory. The great Dutch geophysicist Felix Vening-Meinesz, the discoverer of subduction zones, strongly reinforced Griggs's theory with his measurements of the gravity field at oceanic trenches, but the arguments of these two researchers were not convincing to their colleagues. What is more astonishing is that neither Griggs nor Vening-Meinesz made a connection between their theory of convection currents and continental drift. For them convection currents were permanent and stationary: they existed as descending currents at the edges of continents, but continents and currents had little influence on each other.

Only the great Arthur Holmes had achieved a synthesis. In his work convection currents, continental drift, oceanic ridges, and the formation of folded mountains were all integrated into a coherent scheme. However, in spite of the popular success of his book for students in 1945, he also was unable to convince many people. Postwar geologists preferred analysis to synthesis.

By calling the new synthesis of the 1960s "plate tectonics," the new conquerors, most of whom were geophysicists, undoubtedly wanted both to break out of the analytical straitjacket that restrained the geologists and to appropriate to themselves the preeminence bestowed by the word *tectonic*. There is no doubt, also, that some of the negative reaction on the part of geologists can be explained by their resistance to this ambition and semantic appropriation.

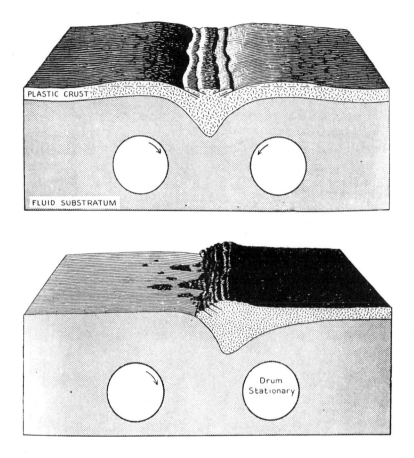

FIGURE 70 David Griggs designed a mechanical scale-model to simulate the action of convection currents on the overlying crustal layers by means of rotating drums. The materials of the "crust" and the "mantle" have properties such as strength and viscosity commensurate with the small size of the model and the short time of the experiments. *Top:* Sketch of the model in which convergent currents have produced a marked downfolding of the "crust." Outward thrusting has developed near the surface in response to the inward drag of the crustal material imposed by the rotation of the drums. *Bottom:* Detail of the result when only one drum was rotated, showing the capacity of a subcrustal current to sweep the crust toward the stationary continent, which is correspondingly thickened. Outward thrust near the surface can again be seen.

Mountain Ranges in the Framework of Plate Tectonics

The first efforts to reconcile tectonics and plate tectonics were made by John Dewey and John Bird, and also Robert Dietz and Warren Hamilton. Their arguments are as follows.

Dietz subscribed to the geosynclinal hypothesis. Taking as his example the Appalachians, as Hall, Dana, and Marshall Kay had done, Dietz suggested that the thrust of the Atlantic "oceanic plate" caused the folding of the sediments that had accumulated in the trough, creating the range that borders the American continental plateau. Although Dietz's explanation was a combination, a synthesis of the theory of geosynclines and that of plate tectonics, it did not achieve much success, particularly among European geologists, who considered it ad hoc and provincial.

Dewey and Bird's contribution avoided this criticism. It was more innovative, and its examples were much more cosmopolitan. It was in fact the first attempt at modern comparative tectonics. John Bird, a professor at the State University of New York at Albany (coincidentally James Hall, the father of geosynclines, had worked in the same city), was not satisfied with Marshall Kay's theory of the formation of the Appalachians because it evaded the question of causes. John Dewey was led to an interest in drift when he realized that the continuation of England's Caledonian range should be sought across the Atlantic. While at Cambridge, England, he was on the fringe of Edward Bullard's group—Vine, McKenzie, Matthews—that was developing plate-tectonic concepts, and he wanted to participate in the tectonic revolution. Hearing the geophysicists speak about mountain ranges without having the least concept of tectonics, he was convinced that a specialist in tectonics could contribute a lot to the new synthesis.

Both Bird and Dewey were trying to understand their favorite mountain ranges. The Appalachians and the Caledonians are homologous on opposite sides of the Atlantic, and the theory of continental drift made them the continuation of each other. Such ideas were at the root of a meeting organized by Marshall Kay at Gander, in Newfoundland, where the two geologists met for the first time. Soon their discussion and collaboration focused on a more general problem than reconstructing the puzzle of the North Atlantic: how to explain mountain ranges according to the new theory of plate tectonics. In May 1970 they published "Mountain Belts and the New Global Tectonics" in the *Journal of Geophysical Research*.

The starting point was simple. The young mountain ranges on

the earth's surface are distributed in a circum-Pacific belt and in an east-west range extending from the Alps to the Himalaya. Adopting the new view of the earth, Dewey and Bird noted that the circum-Pacific ranges are located along subduction zones and that the Eurasian ranges mark a series of continental collisions: the Alps formed by the Africa-Europe collision, the Iranian ranges by the Arabia-Asia collision, and the Himalaya by the India-Asia collision. According to plate-tectonic theory activity should take place at plate boundaries, so they decided to connect the observed topography to the two types of continental boundaries: subduction and collision.

They conceded, somewhat as Dietz did, that subduction generally starts from a continental margin of the Atlantic type (without subduction), which is transformed into a Pacific-type margin (with subduction). Sediments accumulate in the trench created in this way, but there is also pressure and therefore folding and tectonic "chipping." In this way pieces of oceanic crust are tectonically mixed with the folded sediments. The classic mountain-range association between ophiolites (vestiges of the oceanic crust) and marine sediments result, mixed together in what is called tectonic mélange. The Appalachians and the Andes are examples of mountain ranges formed in this way.

According to Dewey and Bird, the second type of range is formed at a continental collision. Continents cannot be swallowed up by the mantle, so subduction ceases when a continent is in a position to be engulfed. This is accompanied by collision, which causes folding and therefore gives birth to mountains. Collisions may occur between a continent and an island arc or, the much more spectacular case, between two continents. The Papua New Guinea range north of Australia is an example of mountains formed by the first type of collision, the Himalaya of the second.

Dewey and Bird's article was very well received by geophysicists, who were already convinced of the validity of plate tectonics, and quite badly received by geologists. The geologists' reasons were most honorable: the models proposed by Dewey and Bird did not correspond to the precisely observed facts. Without examining the objections systematically, I will take note of some of them. The direction of inclination in Dewey and Bird's model of Appalachian tectonic structures is backward. The Andes, archetype of a cordilleran belt formed at a continent–subduction zone border, do not contain any ophiolites. The high Himalaya, archetype for the collision type of range, is actually an intracontinental range: the collision took place at the Tsang-po River, north of the mountains.

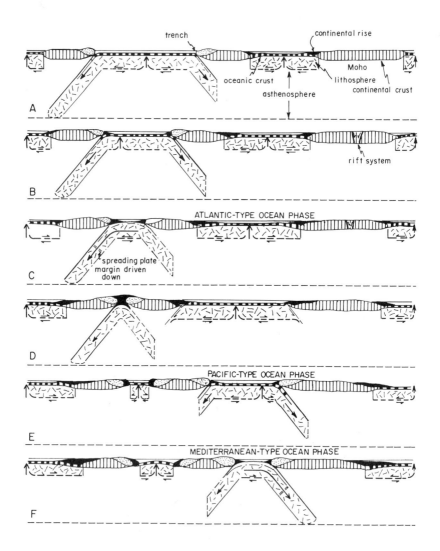

FIGURE 71 Schematic stages in the Wilson cycle of the opening and reclosing of ocean basins (after Dewey and Bird). A Pacific-type basin contracts and an associated Atlantic-type basin opens until finally the Pacific-type basin has completely closed. At this time subduction is initiated at the Atlantic-type basin, transforming it into a Pacific-type basin. The former Pacific-type basin subsequently reopens as an Atlantic-type basin and the cycle repeats. A third type of basin, the Mediterranean, also contracts, but it is unlike the Pacific-type basin in that it is nonspreading.

Scientific theories cannot be constructed on wrong observations!

Nevertheless, Dewey and Bird's article was very important methodologically, because it introduced fundamental concepts that went beyond this or that example. They invited geologists to use their methods and observations to build a new geology. The proposed method consists of linking a defined geologic entity to the site of its geodynamic origin within the framework of plate tectonics. Although this approach was fundamental, it was insufficient to convince a geologic community that was still reticent and, it must be said, stultified by thirty years of the systematic analytic approach. The theory had to be made more flexible before geologists would accept it.

The Tectonics of Asia

The work of Paul Tapponnier and Peter Molnar on Asia must be placed in this context of skepticism and hostility that geologists directed toward plate tectonics. The starting point for both re-

Chang Chen Fa and Paul Tapponnier

searchers was the existence of large earthquakes within plates (rather than at the boundaries), especially in Asia, and the existence of a continental collision near the Himalaya. The earthquake at Tangshan, a city about 100 miles from Beijing, in 1976 caused nearly a million deaths (officially 600,000). China is one of the world's most seismically active countries; earthquakes in that populous land are a real scourge. The first seismograph was invented in China in 1084, and the best seismic annals in the world are found there. The earthquakes, which are both numerous and intense, are located not on a plate boundary but in the interior of the Asian plate. This contradicts one of the postulates of plate tectonics—that seismic energy is dissipated only at the edges of the rigid plates.

In the mid-1970s Tapponnier and Molnar proposed the theory of slip lines to account for these observations. Peter Molnar, a young professor at MIT, was a seismologist who specialized in earthquake focal mechanisms. As a pupil of Isacks, Oliver, and Sykes, he was a product of the plate-tectonic school of Lamont Geological Observatory, and he took plate tectonics as gospel. At that time Paul Tapponnier, a pupil of Maurice Mattauer, was spending a year at MIT studying rock mechanics with Bill Brace. During his stay Tapponnier discovered an analogy between the distribution of great faults in Asia, as they can be seen on satellite photographs, and the slip lines of material motion that are produced during laboratory stamping of rigid-plastic materials such as certain metals or plastics. He therefore proposed that the collision of India with Eurasia should be considered as the stamping of a rigid form—India—in a rigid-plastic medium—Asia. From theoretical models established through laboratory experiments and calculations he determined the distribution of fractures and, what's more, the types of faults to which they must correspond (normal, reverse, or strike-slip).

Molnar then studied the focal mechanisms (which indicate the orientation in which faulting occurred) of earthquakes found on the various faults detected by Tapponnier in satellite photographs. The mechanisms actually corresponded to those predicted by the slip-lines theory.

Molnar and Tapponnier's theory took up, in a more modern and physical form, an idea proposed by Argand in 1930, the existence of plasticity in the Asian landmass. Argand drew flow lines of material and defended the idea that the deep folds influencing the great continental landmass were more important than the spectacular folds affecting the geosynclinal troughs. Translated into modern language, this means that from a tectonic point of view intraplate movements are more important than movements at plate boundaries.

Tapponnier and Molnar revived these ideas, but in so doing they encountered one of the tenets of plate tectonics: the rigidity of plates. Some plates, or at least the continental parts of them, do *not*

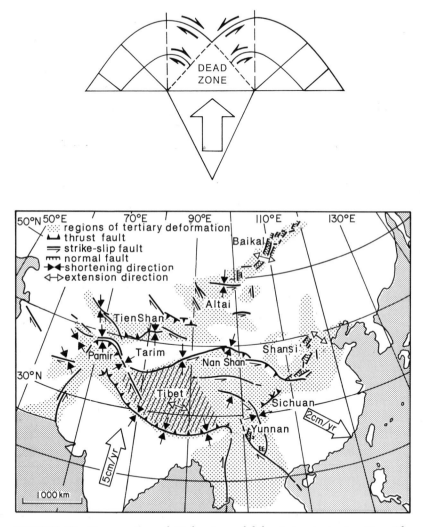

FIGURE 72 Tapponnier and Molnar's model for Asian tectonics. *Top:* The deformation of plastic material intruded by a rigid indenting wedge (or plate). *Bottom:* The application of the model to Asian tectonics. The rigid Indian plate is intruding into a plastically deforming Asian plate.

act in a rigid manner. This principle, which goes far beyond the simple theory of slip lines, encountered very different receptions from different types of readers. Orthodox "platists," like Dan McKenzie or Xavier Le Pichon, were cold to the new ideas, without offering convincing counterarguments except that they were a violation of the fundamental principle of plate tectonics. Molnar, who was close to the brotherhood of plate-tectonic theorists, was upset by this objection and adopted an ambiguous and intermediate scientific attitude, publishing articles based on plate-tectonic principles and on the theory of slip lines. Geologists, on the other hand, fervently welcomed the new ideas. First of all, it must be said, they were so receptive because they were not entirely convinced by plate tectonics, and this new theory vindicated their resistance: an a posteriori justification, surely! But they also advanced valid scientific reasons: for example, how could intra-continental ranges such as the High Atlas of Morocco or the Himalaya be explained if the plates are rigid? What about the great intracontinental faults such as the Altyn-Tagh in China or the Anatolian Fault in Turkey? The existence of some "deformability" within the plates makes it possible to explain the geology of the great mountain ranges of the globe. But then one can turn the postulates around and ask whether all the plates behave in a fragile-rigid way and whether the whole idea of plates must therefore be abandoned, or whether certain plates or parts of plates are rigid only in certain circumstances. No doubt this is going much too far, as we will see.

Like Dewey and Bird's work, Molnar and Tapponnier's articles played a considerable part in the evolution of others' thoughts. In the case of continental geology, it was possible henceforth for people to free themselves from the dogma of plate rigidity.

The most imaginative geologists, whose creative energy had been blocked and inhibited by the intellectual dictatorship of the analysts but who, on the other hand, found the systematic appli-cation of plate-tectonic concepts to geologic facts a bit simplistic, took off full blast. By using the observed facts, the proper domain of geology, they recovered their own creative powers. This effort went well beyond the theory of slip lines, and from it developed a new geology that grew and flourished. The theory of continental mobility restored the driving force to the field of geology that the abandonment of the baked-apple theory had taken away, and plate tectonics gave it a framework that it could accept, but also transcend, from time to time.

Subduction Ranges

Tapponnier and Molnar's work on Asia clarified and extended the model of collision-caused mountain ranges introduced by Dewey and Bird. It was possible to apply these concepts to other collision ranges and to explain the Zagros in Iran, the Franco-Swiss Alps, or Hercynian (late Paleozoic) orogeny in Europe, as well as the Appalachians. The elongated ranges that run all around the Pacific rim, however, are different. As Bird and Dewey had suggested, their folds seemed to result from subduction. Dewey and Bird had found a general explanation; convincing demonstrations of it had yet to be made. As we have seen, this is not the easiest of tasks.

The Andean Cordillera

The Andean range is an elongated mountain belt along the west coast of South America, running from Colombia in the north to Tierra del Fuego in the south. Plate tectonics teaches us that along the length of this range a subduction zone plunges under South America and has no doubt done so for quite a few million years. The region has all the classic characteristics of a subduction zone: a deep oceanic trench running along the coast, deep earthquakes outlining the Benioff-Wadati zone, and some of the largest volcanoes in the world strewn throughout the range. The products of the volcanoes are rich in silica. The most abundant rock found there is called andesite.

The folding of the cordillera dates from the Mesozoic and the Cenozoic eras. The folds are very characteristic of the area. Great overturned strata and thrust sheets are not found there, as they are in the Alps or the Himalaya. Rather, the folds are simple, regular, and often symmetrical. The jumbled, chaotic appearance of collision ranges is here replaced by orderly-looking folds. The rocks seem not to have been subjected to intense thermal phenomena, but all along the range large masses of granite have been intruded, pulling extremely rich metalliferous veins in their wake. The Inca gold that Pizarro carried away came from those veins. The Potosí silver, to which some historians wrongly attributed the European inflation of the sixteenth century, Bolivian tin, Chilean copper—all these riches are connected to the Andean granites.

The Andes is not a minor mountain range! The zonal distribution of its riches shows that its characteristics are not uniform from south to north. In the south, in Chile, the Benioff-Wadati

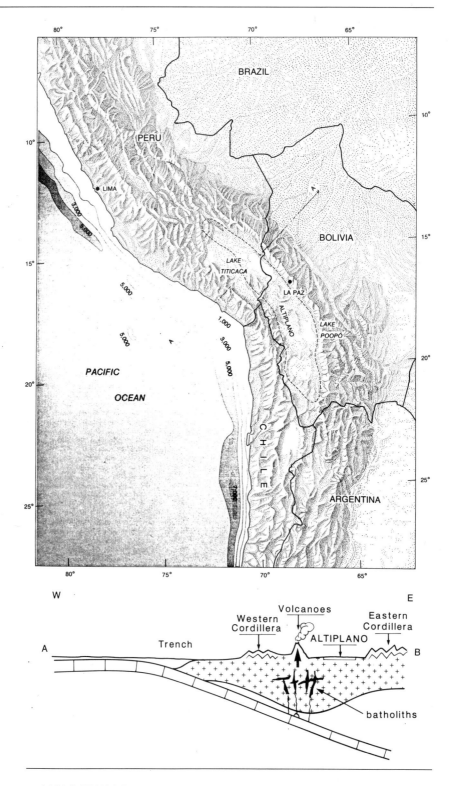

zone is slightly inclined—remember that Uyeda used it as his typical example of a compression zone. The range is more jaggedly folded in the middle region of Peru and Bolivia, where the Benioff-Wadati zone is ill defined, the thickness of the crust is almost doubled, and behind the range there is a plateau at 4,000 meters, the Altiplano, which resembles Tibet in many ways (in the appearance of the inhabitants and the houses, as well!). To the north, in Colombia, the Benioff-Wadati zone becomes well defined again, volcanoes are lacking, and the mountains resemble an alpine range.

In the south and in the extreme north pieces of ancient ocean beds (ophiolites) are mixed into the folded mountains; in the center, in Peru, there is no trace of transplanted ancient ocean floors. How can this range be explained? The basic idea is simple. The subduction of the oceanic lithosphere exerts a force like a battering-ram against the edge of the continent, and this thrust produces accordion pleats. If this is in fact the case, the pleating should be an almost continuous process, but on-the-spot observations show that the folds are temporally limited and that today, while subduction is in full force, a detailed study of Quaternary formations shows that the Andes are in extension, not compression. What does this mean?

Must we admit that the subduction regime is discontinuous and chaotic, with phases of violent acceleration and slow phases? Perhaps there is another phenomenon that we cannot see—perhaps the Altiplano and its thick crust are the product of a collision, like Tibet, but with one of its elements completely buried under the Andes . . . But I shall not attempt to resolve here a debate that is still in progress.

The Mountains of Papua New Guinea

New Guinea is a large island situated to the north of Australia and better known for the customs of some of its primitive tribes than for its geology. In a very difficult logistical context—roads are nonexistent and virgin forest masks most of the outcroppings—Hugh Davies of Australia, the head of that country's geological

FIGURE 73   *Top:* The High Andes in Peru and Bolivia. *Bottom:* The different geologic units of the Andean subduction zone.

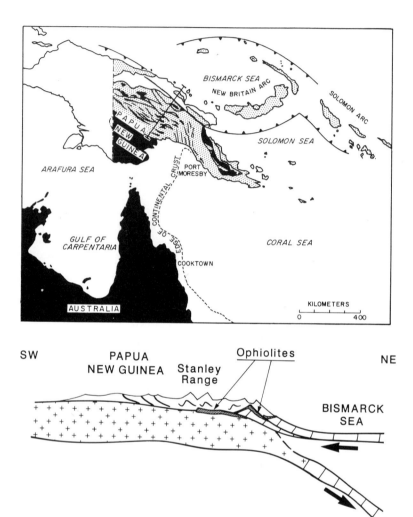

FIGURE 74 Papua New Guinea is an example of a collision between an island arc and a continent. *Top:* The New Britain and Solomon island arcs border the New Guinea–Australian continent. Rock that has escaped deformation is shown in black. *Bottom:* The different geologic units of this convergence zone. Note in particular the presence of ophiolites, or obducted pieces of oceanic crust, that have been pushed on top of the New Guinean continental landmass.

survey, was nevertheless able to accomplish first-rate work. Beginning with a study of ophiolites on the northern coast of Papua New Guinea, Davies established that the period in which the ophiolites were stranded (one says "obducted" in scientific terminology) coincided with the period of folding that gave birth to the great range of mountains that crosses the center of New Guinea (the Stanley range). On the basis of a series of meticulous observations of the unforested coasts, eroded ravines, and bare peaks, all explored by helicopter, he reconstructed the sequence in which the ophiolites were laid down and, at the same time, the origin of the folds.

Let us imagine an ocean-ocean subduction zone, such as the one in the Marianas Arc or in the New Hebrides, that was located north of New Guinea 70 million years ago and diving toward the Pacific Ocean. As time passed, the zone approached the coast of New Guinea, making contact with it 50 million years ago. The trench then filled with sediments. Subduction continued and the continent was pulled under the oceanic crust in such a way that the crust overlapped the continent, smashing, folding, and deforming under it the sediments of the trench as well as the superficial parts of the continent.

This subduction could not continue, because, as we know, the continent was too light to sink into the mantle. After the process of subduction ceased, the continent tended to rise toward the surface, arching its "back" and bringing with it several thousands of cubic meters of fragments of oceanic crust and folded sediments. In this way the heights of New Guinea were formed. This scenario—not a case of continent-continent collision, but of ocean-continent collision—is typical, not unique, and allows us to explain some of the circum-Pacific ranges from New Caledonia to the Philippines by way of part of Japan.

Tectonic Superposition and the History of the Himalaya

Subduction and collision: nature seems to have separated these two types of mountain ranges. But is this really the case? Let's go back to the typical collision range, this time trying to understand its geologic history in light of modern studies, especially those of the Franco-Chinese Program of the Tibet/Himalaya Zone of 1979 to 1983, which engaged more than two hundred specialists from various disciplines. We shall not worry about the details of the reasoning or the data that support it—our objective here is the sequential connection of events.

India broke off from Africa 120 million years ago and began to drift northward. Therefore subduction necessarily began to take place between India and Asia. It swallowed up the entire ocean that had existed in that space. It is known that at first the subduction was of the ocean-ocean type, with an episode of obduction and the laying down of a large mass of ophiolites. Then a subduction directed toward Asia began on the edge of Asia. In the Tibetan region of this continent a situation comparable to that in the Andean Cordillera developed—low-intensity folding, copious magmatism, and volcanism.

The subduction continued and India came into contact with Asia 45 million years ago, with intense folding and the development of major overthrust structures. Then the subduction became blocked, but it continued in Indonesia, creating considerable torque on the Indian plate. The lack of symmetry of forces led to a break in the Indian plate 25 million years ago. A great overlapping of the northern part on the southern part began, and the Himalayan range was born.

Contrary to what is often said, this range is not situated on the seam between India and Asia, where the two landmasses are welded together, but in the interior of the Indian block. Clearly the highest mountain range in the world is a consequence of the collision, not its leading edge. The suture is found farther north, in the valley crossed by the Tsang-po River, the Tibetan name for India's Brahmaputra. And to its north lies fabled Tibet, a plateau 5,000 meters above sea level having a crust 70 kilometers thick, recently raised to this great height.

The interesting part of reconstructing the history of this region is the unfolding of events. In less than 100 million years a mountain range with obduction of ophiolites, an Andean subduction range, a continent-continent collision, and an intracontinental range all developed in succession. The structure that we see today is a superposition of structures, not a pure type.

This is not an isolated example. The great Cenozoic collision between Africa and Europe that led to the formation of the Alps was also a sequential evolution. In that case an ocean-continent collision was followed by a continent-continent collision. The Appalachians, contrary to earlier beliefs, are the product of a collision between Africa and America in the Paleozoic era, after which the range passed through phases analogous to those of the Alps. The Rocky Mountains have a much more complex history in which subduction, collision, and subduction occurred in succession.

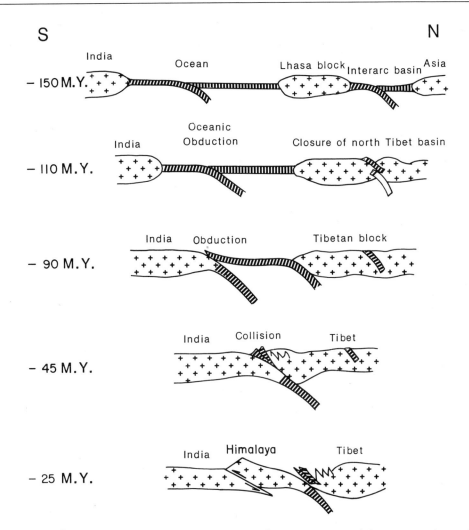

S   N

India   Ocean   Lhasa block Interarc basin Asia
– 150 M.Y.

                        Oceanic
India   Obduction   Closure of north Tibet basin
– 110 M.Y.

         India   Obduction   Tibetan block
– 90 M.Y.

              India   Collision   Tibet
– 45 M.Y.

              India   Himalaya   Tibet
– 25 M.Y.

FIGURE 75   Model of the formation of the Himalayan mountains. (The events prior to 90 M.Y.B.P. are still being debated and are somewhat controversial.)

Tectonic Variety and Complex Tectonics

Many mountain ranges cannot be classified as a single type but must instead be seen as the superposition in the same place of *several* successive mechanisms. Obviously, various combinations of the mechanisms of subduction and collision can create a myriad of structures. Therefore it is necessary to discard the old, inflexible classification of mountain ranges.

The basic concept is that of *variety*. The idea of a single mechanism and a single type for explaining or classifying mountain ranges is dead. Each range is singular and specific. But the observed diversity, the varied scenarios, seem always to result in the combination, in space and time, of two fundamental mechanisms: collision, either ocean-continent or continent-continent, and subduction. Complexity is the result of the combination of these two simple mechanisms under variable geodynamic conditions. To this complexity we must add the variety that results from the particular forms of a specific geography—the shape of the coasts, the orientation of a subduction zone, and so on. Nature offers us an infinite series of geometric variations.

This new point of view gave both freedom and unity back to geology. Research into basic mechanisms became the goal once more. Scientists tried to distinguish among natural examples: where andesitic volcanism is found in an ocean-ocean subduction zone, as opposed to the Andean ocean-continent case, for example, or where an ocean-ocean obduction is not disturbed by the tectonics of a large continent-continent collision. The mobility theory supplied a framework for tectonics that allowed geologists to leave the narrow confines of the analysts' point of view and the overly rigid rules of the tectonophysicists. The danger of analyzing such a variety of cases is that geology may fall back into the trap of compiling regional catalogues!

In previous chapters we have seen two fundamental consequences of the revolution in mobility theory: the creation of a new geology of oceans and plate boundaries and the restructuring of the modes of thought of traditional geology—that is, tectonics. In both cases results to date must be considered only rough sketches, the first models. It is probable that studies currently being undertaken will lead to a better understanding of basic mechanisms. It is just as probable that much of what is considered plausible today will be outmoded tomorrow. But the method will remain: accepting the diversity of the world as it is, but attempting to uncover the basic mechanisms whose combination ensures this variety. In this the mind-set of modern geology parallels that of modern biology.

THE CONTINENTAL CRUST

8

UNTIL NOW I have focused on continental drift, the mobility of the continents, and the various consequences of this mobility without asking how the insubmersible and mobile continental rafts were formed. What special chemical composition permits continents to float on the surface of the mantle? How did the earth fabricate the material? When did this occur?

The composition of the mantle and that of the terrestrial core are known to us only indirectly, through measurements and complex reasoning. We have no direct, continuous access to the earth's interior. Such is not the case for the continental crust, which is directly accessible to observation. The great faults and deep valleys in mountain ranges allow us to sample thicknesses of more than ten kilometers. Only the deepest parts are not directly accessible to us.

The study of continental rocks for nearly two centuries has taught us that in spite of considerable diversity these rocks have an average chemical composition close to that of granite. Granite is composed of two principal minerals, quartz and feldspar. Compared with the two other rock types—basalt and peridotite—that constitute the oceanic crust and the mantle, respectively, granite is enriched in silicon, aluminum, sodium, and potassium. To indicate this enrichment the continental crust, rich in silicon and aluminum, used to be called SIAL, and the silicon- and magnesium-rich mantle was called SIMA.

For our purpose density is the fundamental physical property among those that distinguish quartz and feldspar, the minerals of the continental crust, from olivine, the mineral of the mantle. Olivine is a heavy mineral; quartz and feldspar are light. Because of this contrast in density the continents float on the surface of the mantle and are not engulfed by it at subduction zones. Unlike the ocean floor, whose life expectancy never exceeds 200 million years, the continents, once formed, last throughout geologic time. The *oceanic crust* is constantly renewed, the eternally young segment of the terrestrial surface, whereas the *continental crust* is ancient. Because they have remained at the surface so long, the continents contain information about the long history of the earth. They are the earth's archives. Without the fundamental property of flotation of the chemical constituents of the continental crust, we would have no archives from which to reconstruct history. The earth's memory would be limited to only the last 200 million years!

Any approach to the fundamental processes that regulate the activity of the terrestrial globe would therefore be incomplete if it did not attempt to explain how this archive was formed, how the thin pieces of shell that constitute the continental crust were differentiated from underlying layers. It is logical, then, to regard the earth as an immense chemical factory that transports, transforms, modifies, mixes, and differentiates a series of complex chemical compounds, called silicates, whose common element is silicon. As in organic chemistry all biological evolution is dominated by the chemistry of carbon and its compounds, earth chemistry is dominated by silicon and the compounds that it forms with the ten other chemical elements that are abundant on our planet, most especially with the most abundant of them, oxygen. Only by understanding this chemistry can we reach an understanding of the intrinsic mechanisms that regulate all terrestrial evolution.

The Chemistry of Silicates

All the properties of the mantle and of the crust are thus dominated by the chemistry of silicates. Let us look at their "molecular" structure.

Matter is composed of atoms. Atoms unite to form molecules and crystalline structures. These molecular units, replicated in the thousands, constitute the macroscopic objects that we see. The physical properties of these objects (hardness, color, density, form) depend on their internal structure, on the way their constituent atoms are grouped. One of the aims of chemistry is to explain the macroscopic properties of minerals or molecules by examining structures existing on the atomic level. Thus we study the atomic structure of silicates to understand the properties of these minerals.

Silicon can join with four atoms of oxygen to form a tetrahedral structure (just as carbon forms tetrahedrals by joining with four hydrogen atoms) with the silicon in the center and the four oxygen atoms on the edges. Imagine three Ping-Pong balls lying in a triangle on a table with the fourth above them and in the middle,

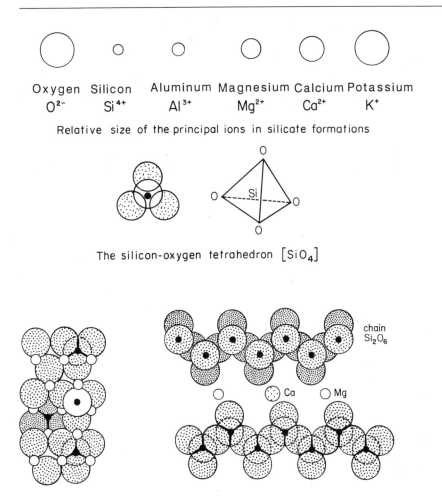

Oxygen   Silicon   Aluminum   Magnesium   Calcium   Potassium

O^{2-} Si^{4+} Al^{3+} Mg^{2+} Ca^{2+} K^+

Relative size of the principal ions in silicate formations

The silicon-oxygen tetrahedron $[SiO_4]$

chain Si_2O_6

Ca   Mg

Olivine   Pyroxene

FIGURE 76   The principal ions found in silicate formations vary in size and charge. These ions combine with the elementary silicon-oxygen tetrahedron (SiO_4), which can be polymerized or linked into more complex structural units. Two common silicates formed of SiO_4 tetrahedra are shown here, olivine and pyroxene (diopside). Olivine is considerably more compact and denser than pyroxene.

and you will have the form of the four oxygens of a silicon-oxygen tetrahedron (SiO_4). The silicon, which is tiny but an indispensable cement for this liaison, is found in the center, in the space left free by the four spheres of oxygen. SiO_4 tetrahedra, like methane tetrahedra (CH_4), are able to polymerize, that is, to link with other molecules to form complex assemblages containing a large number of tetrahedra. The polymerization of SiO_4 tetrahedra is the basis of earth chemistry.

The dimensions of the atoms of silicon and oxygen are very different. The oxygen atom is six times larger than that of silicon. Silicated polymers therefore appear to be assemblages of large spheres of oxygen whose stability is assured by the central atoms of silicon. Actually, the SiO_4 assemblage is not electrically neutral but carries an excess of negative charge $(4-)$; it is a complex ion. In the same way, polymers of SiO_4 are not neutral; they are ionic macromolecules. The electrical neutrality of the whole is achieved by the presence of positively charged ions, or *cations*. (Negatively charged ions are called *anions*.) Cations are lodged in the holes left open among the large oxygen atoms and neutralize the structure electrically. In natural silicates the cations that can play this role are the ones that are sufficiently abundant on earth. They come from the light elements sodium (Na), magnesium (Mg), aluminum (Al), potassium (K), and calcium (Ca) and the heavy element iron (Fe), which, for reasons of nuclear stability, is very abundant in the universe.

Cations have different sizes and charges. For example, the ion Na^+ has one positive charge and the ion Ca^{2+} has two, but the ions are of similar size. Fe^{2+} and Ca^{2+} have two positive charges each but differ totally in size. The "fit" of the cations with the framework of silicates depends on a congruence of both charge and size. Thus small ions such as magnesium and iron can fit into very compact silicate structures in which all the oxygen atoms are tightly joined together. This compact structure—such as the associations of tightly packed oxygen anions and small cations of magnesium and iron that make up the minerals *olivine* and *pyroxene*—results in a high density. In contrast, large cations like potassium and sodium can bond with silicated polymers only if the oxygen anions are not too close together, that is, if the spacing of the oxygens leaves "holes" large enough to accommodate them. This is the case with feldspar, whose loose structure gives it a low density. The composition of the mantle is dominated by olivine, a dense mineral rich in magnesium and iron; that of the continental crust by feldspar, a lighter mineral with a high potassium and sodium content.

We already know that the rocks of the mantle behave like fluids in the span of geologic time and space appropriate to them. If a lighter material is poured onto the surface of a fluid, it will float. Similarly, the continental crust, which is light in comparison with the mantle, floats on it. The density of the upper mantle is 3.2 grams per cubic centimeter, and that of the rocks of the continental crust is 2.8. As I pointed out in the discussion of isostasy in Chapter 1, one can easily calculate the "equilibrium" thickness of the continental crust by using Archimedes' principle. These calculations show the thickness to be 30 kilometers, about the same depth that seismologists have found for the Moho. The idea of continents floating on the mantle, so dear to Alfred Wegener, has a foundation in the principles of chemistry.

Everything happens as if, over the course of time, the mantle were getting rid of "intruders," ejecting toward the surface the large ions of potassium, sodium, and aluminum that cannot enter into the compact structure of olivine. The consequence of this ejection is that molecular structures loose enough to accept the large ions are formed on the earth's surface. These structures are therefore light and capable of floating on the mantle.

It remains to be seen *how* the light continental crust, predominantly quartz and feldspar, is differentiated from the denser mantle, whose composition is mostly peridotitic and whose dominant material is olivine. How is such a stew concocted? As always in geology, this question has two facets: how and when.

Let us abandon for an instant the atomic scale of molecules and crystals, spanning but a few hundred angstroms, for that of the continents, which cover thousands of kilometers!

Orogenic Belts

Let us look at the continents as the geologic map of the world depicts them for us. In our mind's eye let us erase the thin skin of horizontal sediments that covers some of the superficial parts of basins, such as those of Paris, Moscow, London, the Amazon, Michigan, or Texas. The continents then appear to be made of strongly folded rocks into which igneous rocks such as granite are intruded. Dates for both folding and granitization can be determined, and the formations corresponding to a given geologic epoch can be mapped to define provinces several thousand square meters in area. In North America these include a recent province (less than 200 million years old) that contains the western cordilleras, a Hercynian province that contains the Appalachian range, and a series of Precambrian provinces that contains the Canadian shield and the subbasement of the Great Plains region.

Such a map of the world shows that the provinces that are more than 1.6 billion years old occupy the centers of the major conti-

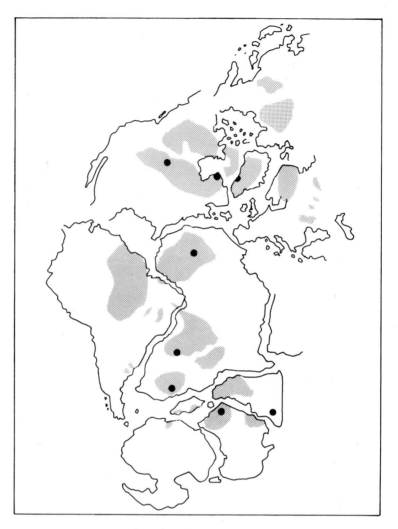

FIGURE 77 Continental nuclei older than 2.5 billion years are shaded on this reconstruction of Pangaea.

nents. This becomes even clearer if the continents are shown in the positions that they occupied before the breakup of Pangaea. The more recent ranges, and in particular the Paleozoic, Mesozoic, and Cenozoic ranges, are located around the periphery of these old nuclei, at the edges of the Precambrian shields. In the mountain ranges of southern Europe the folds are less than 140 million years old. In middle Europe, from Brittany to Bohemia, the folds are 280 to 480 million years old: this is Hercynian and Caledonian Europe. Much of northern Europe, Scotland, and Scandinavia consist of ancient Precambrian formations that are 1, 1.6, and 2.7 billion years old.

The formations seem to be arranged as if younger and younger orogenic belts had gathered around an ancient Scandinavian core. But this arrangement is not a recent phenomenon. When the provinces are defined according to their geologic ages, an analogous zonality is seen to exist, even in the interiors of Precambrian shields. Al Engel of the University of California at San Diego has shown that the North American shield consisted of a nucleus 3.2–2.7 billion years old, called the Superior Province, which was

FIGURE 78   Three essential tectonic assemblages in Europe: the Scandinavian shield (ages in billion years are given); the Caledonian and Hercynian mountain belts, formed between 450 and 250 million years ago; and the Alpine zone of southern Europe, formed in the past 150 million years.

encircled by the 1.8-billion-year-old Churchill Province, and then by the Grenville Province, whose formations are 0.9–1.2 billion years old. An analogous zonality has been shown for Precambrian formations in Scandinavia, Australia, and South Africa.

The belts that mold themselves successively around a central nucleus and enlarge its surface are *belts of folded rocks*. Here we reencounter the idea of geosynclines, troughs *bordering* continents whose future is to be folded into mountain ranges. By wearing down their summits, erosion allows us to penetrate to the very heart of mountains, and we have learned that the formation of a range is not simply a question of superficial folding. It also involves heat transfer, metamorphism, melting, and the formation of granites. Remember, now, that the continental crust is composed of metamorphic rocks and granites. *From this observation came the idea that the underlying function of orogeny, besides that of creating imposing heights, is the creation of pieces of continents.* Thus the successive episodes of orogeny that accrete around ancient nuclei are successive increments of continents. *Orogeny is therefore the genetics of the continents.* The formation of the continents and orogeny are intimately related processes. Furthermore, because orogenic episodes have succeeded one another, the surface of the continents must have increased over geologic time. The continuous growth of continental surface points to an early period when earth was uniquely "oceanic," and also to a future in which continents, like waterlilies on the surface of a pond, will invade the entire area of the oceans. But if this is the case, how were they able to grow, and at what cost? The only possible source of continental material is the mantle, so the continents must have been created continuously at the expense of the mantle. But how did such a phenomenon take place?

On the geologic time scale, I have already reviewed the possible role in this process of volcanism in subduction zones. On the atomic level, I have pointed out that the formation of continents consists of an expulsion from the mantle toward the surface of large ions such as potassium, sodium, and aluminum. But did this expulsion take place in a continuous way, over the course of geologic time, or only once, in the very distant past?

The Growth of the Continents

These questions still arouse passionate debate. To focus our discussion, I will examine two opposing arguments.

The argument that continental crust remains constant in size.

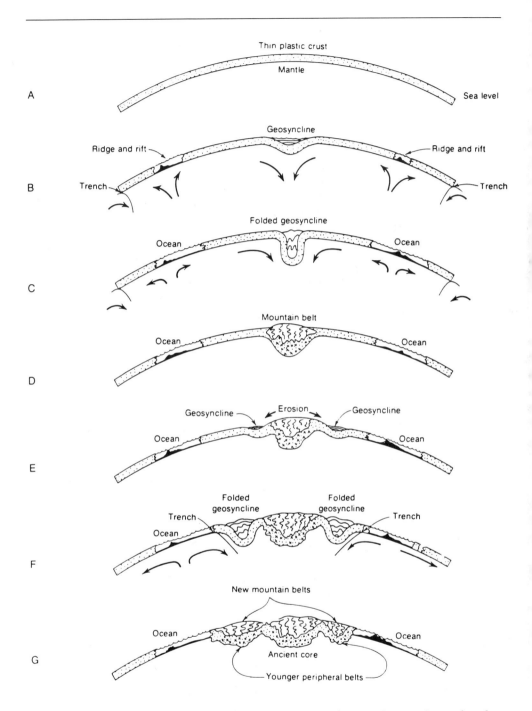

FIGURE 79 A unified model of plate tectonics, geosynclinal theories (note the close similarity to Griggs's model), and continental growth surrounding older continental core units.

The growth of the continents over the course of geologic time is only an appearance, an illusion. In fact, the continental material was differentiated during the first moments of the differentiation of the earth into concentric layers: core, mantle, crust, hydrosphere, atmosphere. Since that time, the same continental material has been used time and again for orogeny; it has been constantly recycled and reworked. The cycle of erosion, sedimentation, and burial in the trenches, and then of metamorphism and remelting of the metamorphic rocks, acts constantly on the same materials that float eternally on earth's surface.

To support this explanation one can refer to observations that have long been made by geologists: under the recent folds of the Alps or the Pyrenees there lies a granitic basement that is much older (Hercynian and Precambrian). In the American west (Arizona) under the 60-million-year-old Laramide range is a subbasement that is 1.6 billion years old. The theory of palingenesis—according to which granite, for example, is formed by the remelting of sediments of continental origin or of pieces of ancient granites from deeper regions—can also be used to support this argument. (I discussed palingenesis in Chapter 2.) It is a tectonic illustration of the continuous recycling of continental materials, of *crustal rejuvenation*, as it is called in technical terms. Episodes of orogeny are merely instances of tinkering with old materials. Mountains, to use an expression the biologist François Jacob coined for genetics, are but new things made from old.

Further support for this argument comes from observations of the successive advances and retreats of oceans onto the continents, which can be deciphered by reading geologic strata. If the surface area of the continents increases, that of the oceans must diminish. Sea levels must rise, and since erosion has brought down the heights of the continents almost to sea level, water must cover the continents. Therefore the old continents should be submerged under the sea! But the Canadian Precambrian shields, as well as the Siberian, Brazilian, Scandinavian, and African ones, show us the contrary. The sea that used to cover Europe, Africa, and part of America, leaving behind ancient sediments containing marine fossils, has receded from these continents! This "proves" that the surface of the continents has remained constant or perhaps even diminished, confirming a stationary view of geologic history. The "actors" were defined at the moment of the "big bang," at the dawn of geologic time, and since then the geologic cycle has unrolled in an immutable procession, constantly renewed but always the same. What we observe today has always taken place.

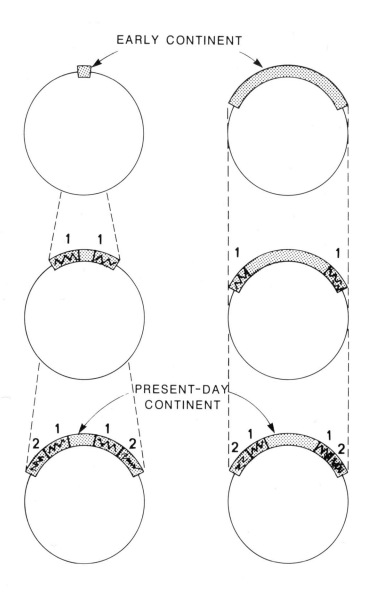

FIGURE 80   Two contradictory theories of continental development. On the right is shown the theory that continental crust was created once and undergoes continual recycling of ancient continental material in convergence zones. On the left is the view that continental material is continually created in convergence zones.

This view is a systematic extrapolation of Charles Lyell's famous principle of uniformitarianism.

The history of the earth is divided into two episodes: the episode of creation, during which formations were put in place and the rules of the game defined, and "afterward," the episode of geologic time during which successive cycles, identical to one another and following the same rules and limiting conditions, unfolded. The study of recent geologic time furnishes the rules for this. All that remains is to extrapolate them further into the past.

The argument that the continental crust increases in size. The continental crust is extruded from the mantle *continuously,* and the amount of continental material grows unceasingly over geologic time.

Continental crust is created at the expense of the mantle purely by magmatic processes. The andesitic volcanism of subduction zones, whose chemical composition is similar to that of the average continental crust, is proof of continual extraction. In this view at least some granites have their origin in the mantle.

Concentric orogenic belts are evidence of the continuous growth of continental surface, which will eventually cover the entire surface of the globe. An argument for continuous continental growth can be taken from the geologic map of the earth, which shows complex formations with varied geologic histories. Rocks of Hercynian age lie under the folded sediments of the Alps; Precambrian nuclei can be detected in some places under the Hercynian formations of the Appalachians. Simply stated, one can define the great geologic provinces by the ages of their dominant folded formations. Given large enough time intervals, such as 450 million years, the proposed division can be made without too much difficulty and the provinces can be mapped, as we have seen. The cartographic extent of each province can be measured, and a graph can be drawn to show the continental surface accumulated in each period of time. Patrick Hurley compiled such a graph and demonstrated an important fact: the total surface area of the continents seems to have increased over time.

Each piece of the present-day continents can be eroded, subjected to a new period of orogeny, and transformed by a new metamorphism at any moment; therefore the ancient continents that are found intact today have survived all threats of destruction. Of course, each time interval brings a certain probability of destruction with it; the older a continent is, the more chances it has had to have been destroyed. Pieces of ancient continents are only relics of much greater entities that once existed. It is possible to calculate

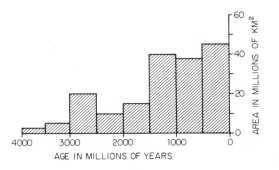

FIGURE 81   Continental area as a function of geologic age (dates determined by fossils and radiometry). (Figure from P. Hurley and Rand.)

the surface area that existed in each period of the past and the rate of continental creation for each epoch. According to Hurley the rate of creation of continental crust has remained constant over the course of geologic time. He provides a simple, quantitative expression for the theory of continental growth.

In this view of things, the earth's surface changes constantly as new pieces of continent are added to the preceding ones. One can therefore imagine that the dynamic flow of the ancient earth, denuded of continents, was quite different from what we observe today. Here a strongly evolutionist view is substituted for the preceding stationary view. As the area of the continents increases continuously, so too the ocean-continent ratio changes in the same way.

As you can see, two totally opposed theories have been built from the same observations and measurements. Can we reconcile them? To explain the flotation of continents we had to descend to the atomic level and make use of the chemistry of silicates. To decide between these two opposed theories, we will be obliged to descend to the subatomic level of isotopes and atomic nuclei.

Isotope Geology

In Chapter 2 I explained that certain long-lived radioactive isotopes, such as uranium 238 and rubidium 87, disintegrate into stable isotopes, in this example to yield lead 206 and strontium 87. Chemical elements such as lead and strontium are found in nature as mixtures of several isotopes, the proportions of which vary because some isotopes are created by radioactivity and others are not. The isotopic composition of strontium or lead taken from different terrestrial rocks is found to vary in the ratio between strontium 87 and strontium 86 and between lead 206 and lead 204. The observed isotopic variations depend on two factors: time and geologic environment.

A very old geologic environment rich in rubidium or in uranium produces rocks with a higher proportion of strontium or lead than is produced in a young environment poor in rubidium or uranium. The isotopic composition of an element containing an isotope produced by radioactive disintegration mirrors the general characteristics of the environment in which it "spent its life." It is a reflection of the element's history.

The isotopic composition of strontium or lead in a rock that has spent only a short time in the continental crust and that is rich in rubidium or uranium is more "abnormal" than that of a rock from the mantle, which is poor in radioactive elements. Isotopic composition, then, provides a method for determining the origin of a rock. For example, are new pieces of continental crust recently formed at the expense of the mantle, or are they recycled pieces of ancient continents? Scientists attempted to answer this question by systematically measuring the isotopic composition of the chemical elements strontium and lead taken from continental, and therefore granitic, rocks. Starting in 1965 a series of studies of this kind was undertaken, most notably by Patrick Hurley of MIT and Clair Patterson of Caltech.

Hurley studied the isotopic variation of strontium in granitic (therefore continental) rocks of various ages and found that their "initial" isotopic composition (that is, their composition at the time of formation) was identical to that of basalts, direct products of the mantle. He concluded that granites are derived directly from the mantle and that the orogenic segments with granitic cores have been continuously differentiated over the course of geologic time.

Clair Patterson was interested in the isotopic variations of lead, which he extracted from samples of high-potassium feldspars of various ages. The isotopic composition of the feldspars led Patterson to conclude that the samples had existed in a reservoir high in uranium (that is, in the continental crust) for a very long time. In contrast to Hurley, he argued that the continental crust differentiated itself from the mantle four billion years ago and that since then it had been reproducing itself from itself.

So the two researchers arrived at diametrically opposed results!

Hurley invited Patterson to spend six months at MIT so that they could try to find a solution to this problem together. The result of this collaboration was not a synthesis, but a break between the two men!

In the late 1960s I began to think that the cause of the difficulty was the use of different isotopic tracers. This led me to propose a model that took account of the two types of observations. In this model, called a mixing model, the continental crust is formed over time through the addition of successive orogenic belts, but each belt consists of two parts: one new part that has just been extracted from the mantle, and one recycled part from the previous continent. Over time the proportion of the recycled part increases to the detriment of the newly formed part. Lead is abundant in crustal rocks and rare in basalts, which are extracted from the mantle. Analysis of lead in a mixture of crust and mantle therefore emphasizes the crustal component. On the other hand, strontium is very abundant in basalts and much less so in granites. Analysis of strontium in a mixture of the two clearly stresses the mantle component. This simple balance-sheet effect explains the difference between Hurley's and Patterson's results. It must be said, however, that for ten years my explanation convinced no one. Everyone was content to accentuate the difference of opinion between Hurley and Patterson.

In 1976 my own group in Paris and then Gerald Wasserburg's at Caltech discovered the existence of variations in the isotopic composition of neodymium in terrestrial rocks. These variations are due to the radioactive decay of samarium 147. The rapid development of this new method of isotopic tracing by the two groups, which were soon joined by Keith O'Nions's Anglo-American group, made it possible for us to show that the only model compatible with the experiments on granites was the crust-mantle mixing model. The use of a new isotopic tracer had convinced the skeptics and supported the mixing model as well!

The success of the mixing theory did not imply that the geologic mechanisms for the combination of recycling and new formation of continental crust had been deciphered. Let me try to outline these mechanisms as I personally conceive them today.

Continental recycling takes place through the well-known geologic cycle of erosion, transport, and sedimentation in the oceanic troughs near the continents (the ancient geosynclines) followed by the transformation of the geosynclines by internal heat to create metamorphic rocks. The contribution from the mantle comes from andesitic volcanism, which I have described several times, notably in relation to subduction zones. The mixture, in other words, has two principal components: the sediments whose chemical composition is a product of the complex processes of erosion by water, a corrosive agent of transport and separation; and the magmatic materials produced by the mantle through partial melting, yielding basaltic products. The whole can be altered, eroded, transported, and sedimented out into new rock layers, themselves intruded by magmatic rocks; the cycle is reproduced in this way.

The intrusion of magma takes place at temperatures of 1,000–1,200 degrees Centigrade. Sediments that are rich in silica and that still contain water start melting at 750 degrees Centigrade. Therefore the intrusion of basic magma (silica-poor material derived from the mantle) causes a remelting of the sediments, the result of which is the formation of granites. Sometimes melting is associated with a mixture of magmatic mantle products and melted sediments, from which intermediate magmas are obtained. On the whole, a new piece of continental crust consists of a mixture of old continental rock and newly extracted material from the mantle.

The segregation of silicon in combination with ions such as potassium and sodium—large ions from which the open and therefore light architecture of feldspars originates—takes place through two geologic processes: volcanism, which enriches the large ions in the magmatic liquid, and the cycles of erosion, transport, and sedimentation, which separate potassium and sodium from smaller cations such as magnesium and iron and, partially, from calcium, which plays an essential role in this scenario. Thus the *water cycle,* symbol and principal agent of the external and visible activity of the planet, is essential for the development of continents. Clearly, melting fueled from the mantle produces basalt and ejects the large ions of potassium, sodium, aluminum, and calcium. This process is therefore an active participant in the segregation of the continental crust, but it does not allow the separation of calcium from the sodium-potassium-aluminum trio, which alone can construct light structures resistant to burying in the mantle. On the other hand, water and the erosion-sedimentation cycle does this very well by separating clays on one hand and limestones on the other.

The earth is a chemical factory in which each agent plays a fundamental and specific role. Without the exterior geodynamic cycle, there would be no sediments and therefore no mixture, no continents and no archives for the earth. The sediments, products of water from the sky, and the magmas, products of the internal fire, unite to give birth to the continental earth on which we live. Sumerian cosmology has been vindicated!

As for the global influence of this phenomenon, suffice it to say that the continents have increased in surface area and in volume in the four billion years since their birth. In recent years it has become possible to define the shape of their growth curve. This was accomplished by using multiple isotopic tracers—strontium, neodymium, and lead—and by modeling their behavior quantitatively. The result was arrived at independently by both Keith O'Nions's Cambridge group and my group in Paris. The growth of the continents is characterized by an almost constant rate and then by a progressive increase until around one billion years ago; since about 500 million years ago, the total mass of the continents has not increased—material is formed but an equal amount is destroyed. Present continental growth appears to be zero; the mass and the surface area of the continents have attained a steady state.

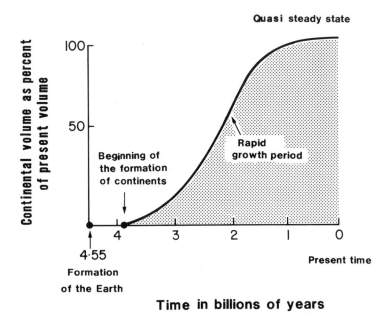

FIGURE 82 Continental growth through geologic time as determined by modelling the evolution of the isotopic ratios of strontium, neodymium, and lead in mantle rocks.

Tectonics and the Growth of Continents

How can a growth curve be related to geodynamic processes? I have recently attempted to establish such a relationship by proposing, with Claude Jaupart, the following model, inspired by the Dewey-Bird interpretation of orogenic belts. It has generally been granted that the creation of new continental surface by the formation of new orogenic belts is related to subduction. The distribution of orogenic belts on the edges of former continents suggests that the phenomena that create the belts are linked to the phenomena of the ocean-continent border. This was the basic idea of the model for geosynclines. But today it is known that geologically active continental margins are those that are linked to subduction. It is also known, as I have stressed in foregoing discussions of subduction zones, that volcanism and plutonism are plentiful in these areas.

If we examine the island arcs such as Indonesia, the Marianas, the Kuriles, or Greece, we certainly find reason to see the precursors of a future orogenic belt in them. The arcs contain significant andesitic and basaltic volcanism, abundant sedimentation of mixed origin (both from continents and from the mantle) whose extreme metamorphism could produce a continental crust, islands whose origin is an older continent and which are relics of it, and oceans that sometimes have continental subbasements and sometimes oceanic ones, a fact that the partisans of the geosynclinal theory used to support it. Imagine that the whole of Southeast Asia were shortened by 500 kilometers and that the Benioff-Wadati zones, which festoon Asia from Kamchatka to Sumatra, were stuck onto the Asiatic continent in the manner of Andean subduction. The island arcs would be folded, metamorphosed, injected with magma, carried along, turned over, and would finally give birth to a vast circum-Pacific orogenic belt whose geologic characteristics would not be very different from those that can be observed in ancient ranges.

Let us concede that the creation of continental surface is related to continental subduction. Subduction zones are linked to the surfaces of continents—or, more exactly, to their perimeters. Would the continents therefore tend to grow indefinitely by this means? We know that they have not increased in surface area much in the last 500 million years. What mechanism regulates this?

The examination of continent-continent collision ranges, and especially of the Himalayan or Iranian ranges, demonstrates a fundamental fact. In the Himalaya, for example, we find a doubling of the thickness of the continental crust from the Indian-Asian suture of the Tsang-po in Tibet to the highest peaks further south. Over the nearly 500-kilometer-wide area, 1,000 kilometers long, the exposed surface is therefore reduced by half. With the help of erosion, in about 30 million years the continental crust will be reduced to its usual thickness of 30 kilometers. This phenomenon will have caused 500 kilometers × 1,000 kilometers × 30 kilometers or 15 million cubic kilometers of continent to disappear!

The continents are unsinkable but not, in fact, indestructible. They escape the violent swallowing into the mantle that continually absorbs the ocean floor, but they suffer the slow, superficial erosion that, along with collision, destroys first continental surface and then the mass of the continents. The sediments produced in this process finally return to the mantle, pulled by the ocean floors that carry them down the subduction planes. The mantle, which is incapable of absorbing continental blocks, assimilates parts of them in the form of sediments. This disappearance of sediments into the mantle by subduction had been suggested by Harry Hess in 1951. Modern results, and in particular the detailed studies of subduction zones made during the underwater drilling program IPOD, confirm this point of view. Because the mantle both creates and destroys the crust the ocean-continent system and, more generally, the internal and external geologic cycles are connected in bidirectional relationships. The components of the exterior cycle (water and the continents) are not only expelled from the mantle, they also influence it in turn by reentering it and modifying it. Realizing this opens a new door in our understanding of geochemical cycles and of the chemical history of our planet.

Thus the growth of continents appears to be the result of an antagonism between the processes that create continents and those that destroy them, between subduction phenomena and collision phenomena. When the earth was young, there were few continents and so collisions were rare. But over the course of time the amount of continental surface increased; it broke into pieces and its numerous and mobile pieces eventually encountered one another and collided. The process of creation of continental surface did not stop 500 million years ago. It is a self-regulating process, the regulating force being continental collision, whose importance increased over time until finally it balanced the creation of continental surface. The continents grew quickly in the beginning, especially before two billion years ago. Then, with the help of collisions, the rate of growth decreased and finally their surface remained constant, as has been the case for the last 500 million years.

Mosaic Tectonics

This model has more general implications for the behavior and the structure of continental units, for the geology of continents. As a result of Franco-Chinese studies in Tibet, I and other researchers have developed these consequences within the framework of a new view of continental tectonics that I call "mosaic tectonics."

First of all, it must be granted that the continents cannot remain large in size for a long time without submitting to internal tensions that break them apart. An example of this is the breakup of Pangaea. Today we know that Africa may be in the process of breaking up along the African lakes, that Asia is crossed by a network of fissures (around Lake Baikal and in the Shansi province of China), and that even France is crossed by deep troughs (the Rhine and the Limagne grabens, for example). These fractures grow by a process called rift propagation, which was well demonstrated by Vincent Courtillot in his study of the Djibouti area (the north end of the African rifting in the Afar triangle—Ethiopia, Somalia, Djibouti). Spreading centers seem to propagate like fissures and in that way fracture the continents.

The paleomagnetic studies of Ted Irving and his former student Jim Briden have shown us that contrary to what Wegener thought, the breakup of continents is not a unidirectional phenomenon that, starting with Pangaea, fractured a series of continents into subcontinents. Rather, episodes of continental breakup and aggregation alternate. In the period before 300 million years ago the earth's surface was covered with pieces of continents that assembled and stuck together to give birth to the supercontinent Pangaea (see the end of Chapter 2), which then broke apart. Later, the collision of India and Asia welded India to Tibet, and the collision of Europe and Africa welded Italy to Europe. This mechanism, which destroys continental surface, also assembles pieces of continents and molds them together. Thus, little by little, the idea that fracturing and welding are two complementary and fundamental processes of continental evolution was born.

Let us try to learn more about the evolution of continents by using paleomagnetism. Carefully studying the formations on the

West Coast of the United States, the Stanford group directed by Allan Cox (who contributed so much to the establishment of the reversal scale of the magnetic field) and Amos Nur noticed that the western American sierras are made of continental fragments or terrains, some of which come from the latitude of Peru (or Australia) and others from that of Alaska. Generalizing from this observation, they postulated the existence of exotic continental terrains in various places around the Pacific. The fragments are of modest dimensions—not like plates half the size of Africa or Asia or even India, but blocks 500 kilometers long by 200 kilometers wide. This led to the idea that the blocks could have been created at the island arcs. Japan, for example, a chain of 3,000 islands, could be moving toward America. This is a novel hypothesis. The classical belief was that island chains were detached pieces of continents (Wegener's bow-and-stern effect) and that eventually they would reassemble around the continent, giving birth to orogenic belts in an accordion-like movement. The theory now is that these continental blocks are unstable and can either be stuck onto the continents again or migrate away from the continental landmass and cover considerable distances. This idea suggests that phenomena of fracturing do not apply only to the great supracontinental fractures such as those that broke up Gondwanaland, but that they also can be created by the phenomenon of subduction itself.

This hypothesis agrees with the one developed for Asia by the Chinese geologist Chang Chen Fa, which has just been confirmed and extended as a result of Franco-Chinese cooperation in Tibet. Chang accomplished this work in the context, which was terrible for him personally, of the Maoist Cultural Revolution, making in this way the most striking demonstration of the failure of barbarity to stifle the creative mind! Let us look at the various elements in his model. The suture between India and Asia, located in the Tsang-po valley, is marked by an ophiolitic belt. Chang noticed that farther north in the Tibet plateau there were other ophiolitic belts that he interpreted as relic sutures. He suggested that there are three relic sutures between Lhasa and the Tarim desert area. From this starting point we postulated that Asia must consist of a series of blocks successively welded to the continent, this being the process of continental growth. This conjecture can be pushed even further by suggesting that *only* these collisions of little continents create mountain ranges; that contrary to what had been thought, subduction creates only modest tectonic formations; and that the Andes range is in fact the result of the hidden collision of

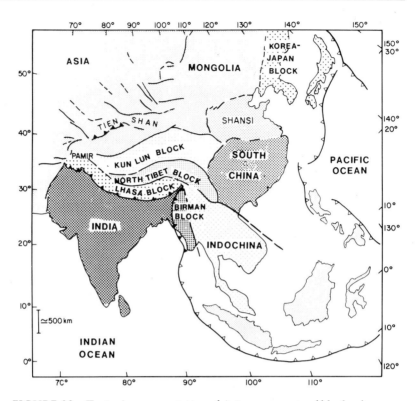

FIGURE 83   Tectonic representation of Asia as a mosaic of blocks that were welded to Asia during episodes of collision.

a microblock that is buried today. This idea is seductive because it suggests all orogenic phenomena are caused by collisions, but it is still in the conjectural stage. To be continued . . .

Thus, little by little, by two totally different routes, the idea was born that the continents are mosaics of blocks of exogenous origin. It is not hard to imagine that if this little game of cutting and pasting has been going on for billions of years it has created a mosaic whose design is almost unfathomable. This explains why it is often so difficult for geologists to reconstruct the development of ancient continents, and why we should not hope to reconstruct ancient puzzles in complete detail! As he digs his way into the past, the geologist must use the *statistical approach* more and more, and at the same time abandon any hope, for ancient terrains, of a cartographic synthesis over vast territories.

The mosaic structure of continents has important tectonic

consequences. Remember Tapponnier and Molnar's attempt to explain the nonrigidity of the Asian continent. If we think of that continent as a mosaic we can understand, as Tapponnier suggested, how the old suture lines between continental blocks can move around when they are given a new push. But they often move in a totally unexpected manner, notably by decoupling from one another and sliding laterally. So a large number of the great strike-slip faults, whether they are transform faults or not, are in fact ancient sutures. Knowing that the great strike-slip faults are seismically active gives us a guide to the reconstruction of continental mosaics, if we accept the idea of dual indicators: ophiolitic belts and intracontinental seismicity. The San Andreas, Jordan, and Altyn-Tagh faults are all ancient sutures.

On this criterion we can divide China, Europe, and the American West into continental microblocks. But we cannot go very far back into the past, because over time sutures "heal," ophiolites erode, seismicity dies down, and the ancient seams become difficult to detect. Moreover, each block defined by this approach is no doubt itself an agglomeration of older blocks, and so on. To these complications must be added the creation of new continental surfaces with accompanying mixing and crustal rejuvenation. That is why we can look into the past only in a statistical way, which is indeed a new way for the geologist!

It is hardly possible to carry these arguments any further at the present time, since they are in no sense resolved. This simple description is designed to show that decoding the continental archives, a task for future generations of geologists, will not be a simple undertaking. The pages in which earth's history have been recorded have been cut, stuck together, torn, and pasted together every which way. Reading them is bound to be difficult!

Do these complex mosaics exist on all the continents, or, more exactly, over the entire surfaces of the continents, or are they restricted to certain areas? There seems to be a certain contradiction between the idea of small mosaics and the simple description of the Canadian or Australian shields, where concentric orogenic belts of more and more recent age have been welded to ancient nuclei. The cartographic picture of an ancient core surrounded by younger orogenic belts seems clear in these regions. Is there some unsuspected complication? Or, on the other hand, must we assume that there are two types of continental landmasses, one obeying the rule of simple belts and the other that of the complex mosaic? Eurasia and western America seem to be the complex continents; the Canadian, African, and Australian shields, the simple ones.

Let us not cut off this debate prematurely, because data are still lacking. Identifying a problem is in itself an important step in science.

Continental Basins

Our picture of the continents is now nearly complete. They are pieces of SIAL extracted from the mantle over the course of geologic time near subduction zones. These pieces attached themselves to older continents. The more complex phenomena of fracture, drift, collision, and welding of the continental rafts occur in concert with the phenomenon of accretion. Thus all the continents were assembled from the folded formations of ancient mountain ranges, more or less leveled off and eroded, depending on their ages. If one looks at the geologic map of the world, however, one finds here and there vast stretches of terrain that are basically circular in form and that do not consist of folded formations. These are the sedimentary basins, like the Paris, London, Moscow, Amazon, Michigan, and Kansas basins. They are great bowls or depressions filled with almost horizontal sediment layers. In the center the sediments are thicker than at the edges. The sediments are not more than 300 million years old, often less. Drilling under the sedimentary rocks has shown that the subbasements consist of folded rocks, of granites and gneiss—that is, of ancient mountain ranges. So the basements of these basins do not consist of anything unusual—like all pieces of continents, they are made of old mountain ranges that have been folded, broken, eroded, and flattened out—but their being covered by young horizontal sediments attests to a peculiar phenomenon. The sea must have invaded wide expanses of the continents and left behind these curious layers of sediments 3,000 to 5,000 meters thick in which one of the riches of the modern world, petroleum, is buried.

How were these basins formed? One hypothesis is that a worldwide force in some periods caused the sea to invade the continents, leaving only a few continental islands here and there. In the Cretaceous, for example, the sea invaded the greater part of France, Africa, and North America, leaving the Central Massif, Armorica, and the Pyrenees in Europe, the Ahaggar Mountains in Africa, and the Canadian shield and the Appalachians in America above sea level. What is the cause of such flooding? Is it climatic? A sudden heat spell may have melted the polar caps, raising sea level nearly 100 meters. Although this answer seems possible for the rapid glaciations and retreats of the Quaternary period, at present we hesitate to apply it to the great transgressions.

The theory of plate tectonics led Walter Pitman of Lamont to propose an original explanation based on the variation in ridge spreading rates. As I have stressed, when a ridge is spreading rapidly its slopes extend laterally over large distances, because young, shallow seafloor extends a large distance away from the ridge axis. This is true in the case of the East Pacific Rise today. When a ridge is spreading slowly, its slopes are steep and its lateral extent small, as in the northern Mid-Atlantic Ridge today. The ridge's axes all reach the same height for equal lengths of ridge, so rapidly spreading ridges occupy a greater submarine volume than slowly spreading ridges, because of their large area of young (hotter, and therefore shallower) seafloor.

Let us imagine that under the influence of a mysterious fever the spreading rate of all the ridges in the world increases. The volume occupied by the oceanic ridges increases and swells up, pushing the water upward and causing the oceans to invade the continents. On the other hand, a period of slow global spreading rates corresponds to a lowering of sea level, a retreat of the waters, a regression. For Pitman the invasions and retreats of the ocean simply express the pulses in the earth's internal history; they are convulsions in the spreading of the seafloor.

This theory is attractive and has aroused a lot of interest, but it is far from universally accepted today, for the geologic and geomagnetic evidence is not conclusive. Skeptics assert that on a global scale the spreading rate is approximately constant, that the acceleration of a ridge in one place is compensated for by the slowing down of a ridge somewhere else. It is as if the earth resists the global pulses suggested by Pitman. But if the theories of climatic variation and of variations in spreading rates are wrong, what alternatives exist? Without wanting to propose an alternative, I will merely point out that the periods of growth of continental surface must necessarily correspond to an elevation in sea level: if continental surface increases, that of the ocean decreases, and since the ocean's volume is constant, depth and therefore sea level must rise. Similarly, periods of decreasing surface and therefore of collision must correspond to decreases in sea level. If this is the case, regression and transgression correspond to collisions and subductions. The tectonic message must be encoded in sedimentation patterns.

Let us leave these as yet unexplained phenomena for the moment and take up the geographically more restricted problem of

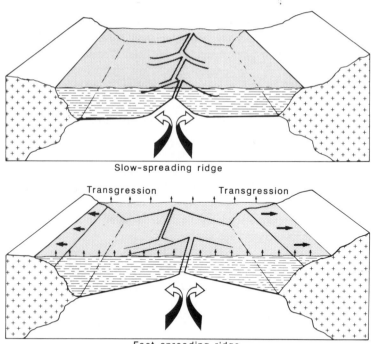

Slow-spreading ridge

Transgression   Transgression

Fast-spreading ridge

FIGURE 84   Pitman and Hayes's theory for sea-level regression and transgression over continental margins. A slowly spreading ocean basin is on
average older and deeper than a fast-spreading ocean basin of the same areal
extent. The shallower ridges and seafloor within a fast-spreading ocean
displace more water and cause the sea level to rise.

the location of basins. Although less grandiose, this problem is no
less fascinating and has economic "fallout" of the greatest importance: a method of finding new petroleum fields.

Why is there a basin in Texas and not on the northern edge of the
Appalachians? Drilling through the sedimentary cover of the Texas
basin has shown that the formations below these basins are almost
identical to those of the Appalachian basin. These formations were
folded one billion years ago. They have the same characteristics,
age, and structure. So? If they were alike 300 million years ago, the
cause of their differences must be sought in more recent history.
The floor of the Texas basin was stretched by great tectonic forces
and heated; it therefore became more plastic and malleable.
Because of the stretching and heating, the floor became thinner and
its altitude decreased. The arrival of water and the deposition of

sediments created an overload, causing the center to continue sinking. Sinking stopped when the basin was full; the basin also cooled, becoming rigid again and therefore ceasing to extend. The problem remains to discover the cause of the stretching and heating. These phenomena were not only at the origin of the opening of the basin, but perhaps also at the origin of petroleum formation by thermal degradation of the organic matter in the sediments.

The issue is far from being decided today, but with the support of the considerable technical means of the petroleum industry, observations and measurements are accumulating and no doubt will help us to learn more. Without getting into an area that is still in full evolution (not to say revolution) and that shows how basic research is stimulating economic applications, I wish to stress two important points. On the one hand, it is easy to conceive that the phenomenon of stretching and thinning of the continental crust could eventually tear a long rent in the crust and cause a new ridge to be born. In that case the formation of basins would be only an aborted break in the crust. On the other hand, one cannot help but notice that the formation of basins again shows us a continental crust that is deformed, deformable, and far from the concepts of rigid plate tectonics. Like intracontinental earthquakes, the origin of basins forces us to see the continents as specific geologic objects. Orthodox plate tectonics regarded them as immobile and passive corks floating along on the conveyor belts of the ocean floor. Modern studies show them to be deformable, breakable, and responsive.

9

THE DYNAMICS OF PLATE MOVEMENT

Plate Movement and Convection

WEGENER tried to find a physical explanation for continental drift; in fact, he lowered his scientific standing in the attempt. No doubt his inability to propose solutions acceptable to physicists and his willingness to present explanatory theories without convincing mechanisms contributed greatly to the rejection of his theory.

Paradoxically, global tectonics has been accepted today without the apparent underpinnings of a causal theory. None of the scientists who developed the concepts of seafloor spreading and plate tectonics attempted to develop a theory explaining the primary causes of drift *at the same time.* It was only *after* acceptance of the phenomenological theory that the scientific community began to research its causes. In no way did this prevent mobility theory from prevailing: the absence of a causal explanation is more acceptable than a wrong explanation!

A closer look, however, reveals some early contributions to the cause of drift. Harry Hess's articles and the writings of Arthur Holmes and W. Jason Morgan included discussions of convection in the mantle. The idea of mantle convection, such a bizarre conception to the physicists of the 1920s (for it was applied to solid matter), gradually made its way into the consciousness of geologists and geophysicists during the 1950s. Of course, it was not officially accepted—the debates between the two great Dutch geodynamicists Rein van Bemmelen and Felix Vening-Meinesz turned out rather to the advantage of the antimobility side—but the mechanism no longer appeared impossible and was in fact considered a promising idea. On the basis of measurements of heat flux or the percentages of radioactive elements in mantle rocks, researchers made theoretical estimates of energy requirements that were compatible with convection of the whole mantle.

When the mobility of the ocean floor was demonstrated in the 1960s, the mobility hypothesis no longer seemed to defy the laws of physics and contemporary physicists (except Sir Harold Jeffreys) did not feel bound to refute it. In contrast to Wegener, the first advocates of seafloor spreading did not attempt to identify the forces capable of moving continents; they merely conceded that these forces were related to the vast movements that activate the mantle. They did attempt to refute Jeffreys's arguments against the Wegenerian hypothesis, especially the objection that continental movements must necessarily be accompanied by folding in the interiors of continents (as when one moves a piece of cigarette paper over the surface of a globe one cannot help creasing and tearing it). To counter this argument the supporters of plate tectonics proposed that the plates are much thicker than the crust and instead are formed of crust plus the topmost hundred kilometers or so of the mantle; they called this mechanical entity the lithosphere. (In effect, they changed the cigarette paper to a piece of cardboard!) These rigid lithospheric plates moved on a much softer layer of the mantle (the asthenosphere), which lubricated their motion.

Using results from experimental petrology, Peter Wyllie, then at the University of Chicago, suggested that mantle material began to melt in the asthenosphere. As I have already mentioned, the melting temperature of peridotites increases with pressure, and therefore with depth. The known facts about heat flux and the experiences of miners show that the temperature increases as one descends into the earth. Wyllie suggested, therefore, that at a certain depth the rock temperature would reach the beginning melting temperature of peridotites. At this point a small quantity of liquid could work its way through the grains of peridotitic minerals, playing the role of a lubricant and making the peridotites plastic.

Laboratory experiments were performed to trace the precise melting point–depth curve of mantle rocks, and the heat flux was measured in order to estimate the temperature-depth curve of the mantle. Drawing the two curves on the same graph, Wyllie showed that their meeting point corresponded approximately to the depth of the transition of the speed of seismic waves (discovered by Beno Gutenberg and studied by Don Anderson at Caltech), which marks the upper limit of the asthenosphere. This zone is characterized by a slowdown in the propagation rate of seismic waves. Above this boundary peridotite is solid and rigid, and wave propagation is good: this layer is the lithosphere. Below it, the peridotite is slightly melted and therefore plastic, so the seismic waves do not propagate well: this is the asthenosphere on which the plates move.

For a long time dividing the mantle into a rigid lithosphere and

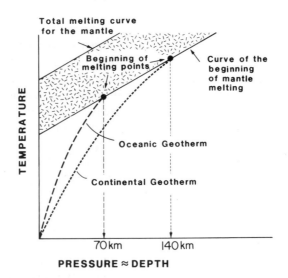

FIGURE 85 The temperature-pressure curves (plotted here as temperature-depth curves) beneath "typical" oceanic and continental regions are superimposed on a pressure-temperature graph. Both temperature-depth curves (or geotherms) intersect the pressure-temperature region where partial melting begins to occur within the upper 150 kilometers of the mantle.

a soft asthenosphere seemed a sufficient theoretical explanation, but it does not explain the origin of the forces that move the plates, nor their distribution. These unknowns were accounted for by reference to "mantle convection," but not much thought was given to the mechanisms involved. The modern period's achievement is to have clarified this concept.

Let me try to explain this difficult problem by approaching it in stages. I will begin with a very simple experiment with which everyone is familiar: the heating of a pot of water on a stove. If you heat the pot a little, you will see no movement in the water—the surface of the liquid is still. When you increase the intensity of the heat you begin to see the surface tremble, and careful observation indicates the existence of currents in the water. If you let bread crumbs fall onto the water's surface, you can see in the middle of the pot an ascending current that divides at the surface and dips down when it reaches the sides. This structure is dynamic, but it is stable and can maintain itself in the same form for hours (assuming that the amount of heat does not change and that the water does not evaporate). Next, turn up the heat. For a while the organized structure remains but then, all of a sudden, it changes.

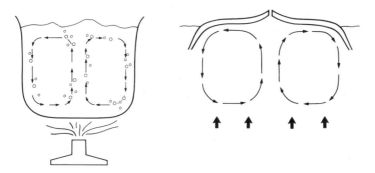

FIGURE 86   *Left:* Convective cells in a pot of water heated from below. Arrows show the convective motions of masses of water. *Right:* Transposition of this familiar picture to plate tectonics and convection in the upper mantle (also heated from below) beneath the plates. The upwelling regions of mantle convection lie beneath the ridge axes and downwelling regions underlie subduction zones.

The motion becomes disorganized and more rapid, continually changing its shape and position. This is the turbulent or disorganized state. If the heat is increased even more, the phenomena of disorder increase, and even more violent phenomena, such as projectiles of boiling water, appear.

The common cause of all these phenomena is the simple fact that hot water is less dense than cold water. A region of hot water isolated within colder water tends to rise, whereas cold water set on the surface of hot water tends to fall to the bottom. This coming and going, the rise of hot water and descent of cold, constitutes the mass transport known as convection.

If the various areas of hot and cold water have time to communicate with one another—that is, to exchange heat—their movement becomes orderly and a stationary, organized structure appears; if they do not, the coming-and-going motion takes place in a disorganized, turbulent way. What determines whether organization is possible is the quantity of heat that can be evacuated by the motion. Mass motion in the pot of water allows heat to be transported from the bottom of the pot toward the surface. If the "load" of heat is reasonable, the transport is calm and organized; if the heat flux is too high, transport is turbulent and anarchic.

The lessons of our experiment apply also to the case of the earth. The analogy is a very rough one, because none of the conditions is exactly comparable, but let us accept the idea that the two convections, on different scales of time and space, obey the same law of fluid mechanics. In the case of the mantle, the source of heat can be the disintegration of radioactive elements in its interior (as proposed by John Joly in the early days of the century) or heat given off by the earth's core.

These heat sources explain how convection may occur in the mantle. The oceanic ridges may be surface manifestations of the ascending mantle currents, subduction zones of descending currents. The distribution of plate edges thus appears to be the most obvious evidence of the convective system. Such is the description that Don Turcotte and Ron Oxburgh, following Holmes and Vening-Meinesz, gave of the spreading of the ocean floor in 1967, even before the theory of plate tectonics was formulated. An important consequence for plate-tectonic theory, that convective movements in the mantle are not stable, arose from the simple observation of the kinematics of plates. We know that the continents move, that the ridges move and change spreading directions, that oceans are born and disappear, and therefore that the distribution of convection currents varies over time. To return to our

Ron Oxburgh and Don Turcotte

pot of water: we must look at it not in its stationary state but in the bubbling state that superheating easily produces, when the motion of the currents is constantly changing, forming complicated and fleeting shapes. The earth's convective state does not consist of a few stationary convective cells. The oceans are not opening and closing with a pendular motion but are transformed and evolve according to schemes whose geometry varies constantly. These are the general considerations on which the research that I will now discuss is based.

Hot Spots

The first theory of plate movement is the hot-spot or hot-plume theory. It was born of the need to explain the genesis of volcanic ranges. In discussing the distribution of volcanic areas along active plate edges, I spoke of volcanism at ridges and subduction zones but I consciously omitted one well-known type of volcanic area: intraoceanic volcanic ranges.

The Hawaiian archipelago is a typical example of this type of range. It extends from the islands of Niihau and Kauai in the west to the island of Hawaii itself, which boasts two majestic active volcanoes, Mauna Loa and Kilauea, and the extinct Mauna Kea. The islands are strung out like a necklace, with the "beads" arranged chronologically. In fact, the Hawaiian islands are just the most visible part of the 6,000-kilometer, mostly underwater Hawaiian Emperor seamount volcanic chain, which stretches from Hawaii to Midway and then veers northwest toward Kamchatka, where it is finally subducted in the Aleutian-Kurile trench. Volcanism on the Midway islands dates to 25 M.Y.B.P., and the island of Hawaii has volcanoes that are still active today. The curve of the decrease in age with distance, presented qualitatively by Reginald Daly but firmly established by Ian McDougall of Australia through potassium-argon dating, is very regular and seems to indicate a well-defined natural phenomenon.

One might think a priori that this chronological progression is the result of the spreading of the ocean floor. In that case the islands would have been created at the ridges and then drifted along the conveyor belt like the rest of the ocean floor. The oldest would therefore be the farthest from the ridges; the youngest, the closest. This is the interpretation that J. Tuzo Wilson gave for the Atlantic islands in the early days of the theory of seafloor spreading, but he soon saw that his idea was erroneous. In terms of our example, the island of Hawaii should be found near the East Pacific

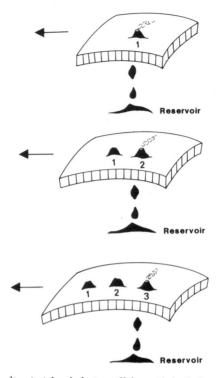

FIGURE 87   A volcanic island chain will form if the lithosphere moves over a fixed region that periodically emits magma. The lithosphere riding over this "hot spot" will be penetrated by volcanic structures that with time form a linear chain.

Rise. In fact, it is 6,000 kilometers away from it. When the ocean floor around Hawaii was dated on the basis of magnetic anomalies (the tectonics of the "zebra skin"), it was found to be Cretaceous (65–130 million years old) and had nothing to do with the adjacent volcanic islands riding on it. Wilson postulated the existence of a hot spot, which is mobile and drifting continuously westward, under the Pacific plate. This hot spot periodically emits plumes of magma that pierce the plate and form a volcano on its surface. As the plate drifts over the hot spot, the magmatic emissions periodically punch through the plate at more or less regular intervals, creating a chain of volcanic islands, exactly like a keypunch operator punching holes in a paper tape. According to Wilson, this hypothesis applies not only to the Hawaiian range but to all oceanic chains. The Tuamotu chain, south of Hawaii, shows the same linear arrangement, and, what is more, a change of direction

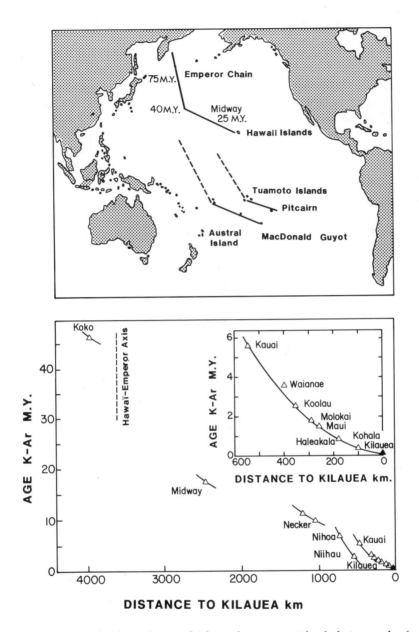

FIGURE 88   Islands in the Pacific form three major island chains, each of which is characterized by an angle where the seafloor is 40 million years old. These chains could be formed by the rigid motion of the Pacific plate over three fixed hot spots. Age-distance relationships in the Hawaiian chain support the hot-spot theory of island chain formation.

can be seen in both chains. For Hawaii, the older segment, which corresponds to the chain of Emperor seamounts, is oriented northwest-southeast, while the recent volcanic chain is oriented east-west, showing that the Pacific plate changed direction of movement about 40 million years ago. These volcanic chains also exist on continents; for those who support the hot-spot theory, the Yellowstone–Snake River volcanic area would be a continental equivalent of the Hawaiian chain, the trail of a hot spot. The volcanic range of the Cameroons extends over ocean and continent, from the islands of Annobón, São Tomé, Príncipe, and Fernando Po to the volcanoes of the northern Cameroons. In spite of the transition from sea to land, the volcanism of the range has the same characteristics all along the way.

The hot-spot hypothesis became influential through the work of W. Jason Morgan, father of plate tectonics. Morgan believed hot spots were formed not directly under the plate but deep in the mantle or even at the mantle-core interface. He also noticed that the relative position of the hot spots on the globe has not varied over the past 50 million years. He therefore suggested that hot spots can serve as a fixed frame of reference, like a series of rigid stakes penetrating the mantle. Thus, according to Morgan, the movements of plates can henceforth be described not just in a relative sense—one in comparison with the others—but in an absolute sense, in reference to a fixed point on the mantle: the Nazca plate is moving faster than the African or Indian plate, for example; the Asian plate is almost stationary; and so on.

At the same time, however, Morgan suggested that the ascending currents coming from deep in the mantle, whose position seems to be stationary over time, are the driving force of mantle convection and therefore of plate tectonics. The currents emerge at points under the continents, causing the continents to split apart. The ridge constructed in this way is maintained or at least directed by periodic injections from deep within the mantle. In this sense volcanic ranges are the surface scars of a phenomenon whose origin is deep within the earth and which governs terrestrial geodynamics. In terms of our experiment with the pot of boiling water, this is the stage at which the hot jets rising from the bottom of the pot suddenly burst through the surface.

To bolster his argument Morgan stressed the fundamental difference between the chemical and isotopic compositions of the basalts of the oceanic islands and those of the ridges, showing chemically different sources for the two types. The idea of two different mantles, one near the surface and the other deep, was later put to very good use by Jean-Guy Schilling, working at the University of Rhode Island. Using geochemical methods to map the northern Mid-Atlantic Ridge, Schilling discovered that the chemical composition of its basalts varies considerably. He found the ratio of some chemical elements—lanthanum (La) and samarium (Sm), for example—to vary by a factor of ten. Moreover, the variations are not random. The ratio of lanthanum to samarium is high near Iceland, decreases in a regular fashion southward, passes through a minimum, and then increases until it attains a value comparable to the Icelandic values near the Azores. Schilling also found that the Icelandic values are similar to values typical of oceanic islands, such as Hawaii, whereas the minimum values, those measured on the ridge near 35° North, are typical of oceanic ridges far from any island. He suggested therefore that Iceland and the section of the ridge near the Azores are sitting on top of two hot spots injected (or trapped) at the ridge. The observed variations are the result of mixing between a hot-spot source and a normal

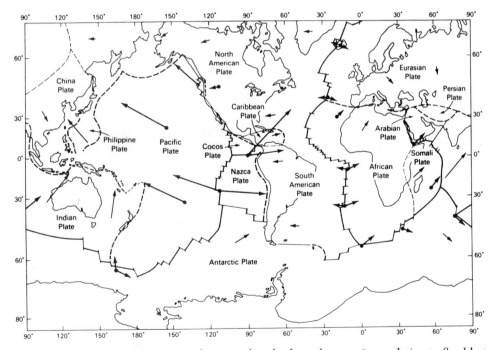

FIGURE 89   Jason Morgan's map of present-day absolute plate motions relative to fixed hot spots. The length of each arrow showing the directions of plate motion is proportional to absolute plate speed.

oceanic source. These studies strongly supported Morgan's ideas, according to which hot spots supplying the ridges are the driving force, or at least one of the driving forces, of plate tectonics.

Starting around 1975, a scattered but luxuriant scientific activity of highly variable quality developed around the idea of hot spots. Some seismologists claimed to have detected anomalies in the rate of propagation of seismic waves under the Hawaiian Islands. They were quickly contradicted. Others tried to show that hot spots induce intraplate tectonics. The existence of a swelling from the mantle below can in effect cause a bulge in the crust that generates extension forces. This idea may have some validity for Hawaii, where volcanism appears to be controlled by faulting, but it must be said that on the whole these demonstrations do not appear to be convincing. By the late 1970s hot spots had become the "hot" theory and they were used to explain everything: plate tectonics, the source of acidic and basaltic magma, and intraplate tectonics. Even the distribution of veins of minerals was related to hot spots!

These ideas are all interesting and perhaps even valid, but at that time they were presented as dogmatic assertions based on doubtful demonstrations. They elicited anti-hot-spot reactions that were hardly more reasonable than the assertions that provoked them. Here again, the scientific method, or rather style of argumentation, is no stranger to this type of confrontation. Scientists often attach more importance to proving or refuting the "proof" of a phenomenon than to the phenomenon itself.

Be that as it may, the debate over hot spots continues to this day.

Are the Plates Pushed, Pulled, or Carried Along?

In contrast to proponents of the hot-spot theory, orthodox mobilists suggested that the solution to the great problem of how the plates move lies in the observation of plates and plate movements. Three possible mechanisms can be distinguished: the driving force is the ridges, which cause the plates to separate; or it is the entire plate, which is carried along by movements in the mantle, especially in the asthenosphere; or it is the sinking of plates at subduction zones, which pulls the plates and disperses them. Mobilists approached the problem by studying relationships among dynamic and geometric characteristics of the plates: between spreading rates (absolute or relative) on one hand and the angle or length of the Benioff-Wadati planes, subduction and ridge lengths, or plate surface on the other hand.

Researchers taking this approach studied earthquake focal mech-

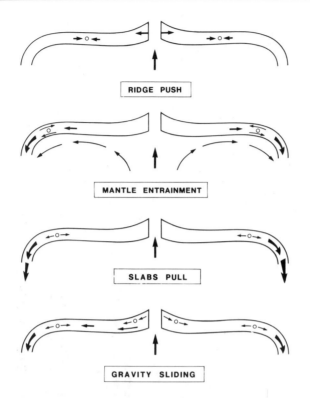

FIGURE 90 The different mechanisms proposed to explain the movement of the plates. Earthquake focal mechanisms have been indicated for compression (→O←) and for extension (←O→). The arrows show the principal forces acting in each case.

anisms in an attempt to describe distributions of stresses within the plates. If the plates are pushed by the ridges, there must be compressive focal mechanisms in the plates, particularly near the ridges; if the plates are pulled by subduction, earthquakes must be extensional; if the plates are pulled along by convection in the mantle, both seismic mechanisms should exist, perhaps with more tension near the ridges and more compression near the subduction zones. The idea that in subduction zones plates are pulled by gravitational sinking into the mantle gained credibility because of Lynn Sykes's study of earthquake focal mechanisms, which showed that except in the immediate vicinity of the ridges intraplate earthquakes are all of the extension type, up to and including those 200 kilometers deep in the subduction zones. It is as if something were "pulling" the plate and subjecting it to extension forces. Other researchers, particularly Donald Forsyth and Seiya Uyeda, showed that the spreading rate and the length of subduction zones of the various plates are correlated. From there it is only one more step to the idea that the subduction zone is responsible for the spreading. The arguments put forward by Sykes and Uyeda lead to similar conclusions and therefore reinforce each other.

Louis Liboutry, a glaciologist from the University of Grenoble, was no doubt one of the first to realize the fundamental role of gravity in plate movement and to make a physically correct analysis of this movement. He showed that for convection phenomena in which the whole fluid is in motion it is dangerous to

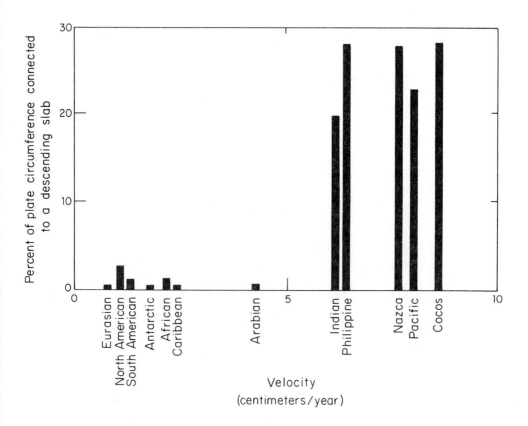

FIGURE 91 The relationship between plate speeds and the percentage of the plate circumference connected to a descending slab. Note the distribution into two families: slow-moving plates with small fractions of subducting plate boundary and fast-moving plates with larger fractions of subducting plate boundary.

THE DYNAMICS OF PLATE MOVEMENT

isolate one part of the system and to attribute to it the role of driving force. Analysis had to be done on the level of the system as a whole. Within this framework he was able to show that gravity was a driving force in several parts of the cycle: ridges, subduction zones, return into the mantle. Seeing plate movement as the driving force for the global system of plate movement means that one must also take into account surface phenomena. In this case the driving force is the rigidity and density of the plates, which causes them to sink into the mantle. As Peter Molnar of MIT suggested, plates cannot sink until they are cold, and therefore dense, enough to plunge into the mantle and create a new subduction zone. The stretching produced by this sinking at the subduction zone would cause the mantle to rise near the ridge zone. This ascent, followed by melting, would set off a new period of spreading and the cycle would continue.

This school of thought, which attaches so much importance to the proper role of plates, does not deny the role of the mantle in global phenomena; rather, it sees it as furnishing the thermal energy that ignites a superficial convective cycle in which the role of subduction is fundamental.

The Mechanics of the "Mantle Fluid"

Although they sometimes feature the same protagonists, the two approaches to plate movement described above—one focusing on hot spots, the other on the plates themselves—must be clearly distinguished from the much more theoretical one based on the mechanics of convective fluids, which states the problem in terms of fundamental physics. The mantle is a convective system. What energy is consumed in it? What are the sources of this energy? What are the physical properties of the "fluid" concerned? And can the primary characteristics of its convection be deduced from the above?

The fundamental physical parameters that must be determined in order to calculate the convective system of a fluid are density, thermal conductivity, and viscosity. In this case, the "fluid" is rather special, since it is a *rocky medium* which, over the scale of geologic time, behaves like a fluid. It is the effect of the time factor that physicists at the beginning of the century did not understand, causing them to reject this physics of the earth's interior, where *solids* behave like *liquids!* Much progress has been made since then, and the concept of a convective mantle no longer seems "physically absurd."

The density of the mantle can be determined directly through seismology. The velocity of propagation of seismic surface waves depends essentially on the density of the materials crossed. Therefore it is possible to "invert" the seismic velocity data and to obtain density profiles from it. The density of the mantle varies from 2 to 5 grams per cubic centimeter.

The viscosity of a fluid measures the ease with which it can flow, losing its shape. A very viscous medium does not lose its shape very much; it does not flow easily. A low-viscosity medium flows and loses its shape easily. The viscosity of the mantle can be determined indirectly, through observations of the modern rise of the Scandinavian shield. I mentioned this phenomenon in the discussion of the theory of isostasy in Chapter 1. Freed from the ice caps that covered it during the Quaternary, the shield rose, causing a local drop in the shorelines. From the speed of the rise Dick Peltier and Larry Cathles calculated the viscosity independently. They obtained figures of 10^{21} to 10^{20} poises. Remember that the viscosities of liquids such as water and oil range from a few hundredths to a few poises. The mantle viscosity is about the same as that of a glass at room temperature (10^{22} poises). The glass we see around us is obviously a solid, but if you enter a medieval church, look carefully at the windows. Usually each piece of glass is a bit thicker at the bottom than at the top. Over the course of a thousand years, it has sagged this minute fraction. This is a tangible demonstration of imperceptible changes in the mantle, which behaves like an elastic solid in the short term but like a viscous liquid over geologic time. Naturally we wanted to know more about this unusual physics!

Volcanoes bring to the surface pieces of rocks snatched from the walls of the media through which they travel. A volcano acts like a natural drilling rig and gives us access to depths of up to 200 kilometers. "Xenoliths" (these wall-rock fragments) are found in many ancient volcanoes, particularly in those that contain veins of diamond. The presence of diamond, which is composed of carbon that is stable only at high pressure, attests to a very deep origin and an extremely rapid ascent for these rocks. Half of the xenoliths brought to the surface by volcanoes are peridotites. These rocks are composed mainly of olivine and pyroxene, and their density is consistent with that which seismologists have measured in the mantle (3.2 to 3.4 grams per cubic centimeter). Their chemical composition, which is very similar to that of the primitive meteorites called chondrites, indicates a high-pressure origin. We also know that the partial melting of such rocks produces basalts. Everything indicates that the mantle is *peridotite* in composition.

Following a series of preliminary studies undertaken in the 1930s by David Griggs and his students at UCLA, researchers began to investigate the mechanisms and circumstances through which the mantle, although made of solid rock, can behave like a fluid on a time scale of millions of years—how it can flow, be deformed, transformed, and finally convected. Working independently, Chris Goetze of MIT and Jean-Paul Poirier of the Institut de Physique du Globe in Paris, a metallurgist converted to the earth sciences, focused on dislocations, or linear defects, in crystalline networks. Displacements are evidence that the mineral has been deformed plastically; in other words, microscopic defects in metals give clues to their macroscopic properties. Goetze and Poirier examined the mineral olivine, its deformations and its structural properties, and their reasoning progressed through the following sequence: microscopic deformation of olivine, macroscopic deformation of peridotite formations, large-scale deformation of the peridotites in the mantle, and mobility of the mantle. The intermediate level of observation—that of the rock or the rock mass—is important, because it can be used to clarify and to test hypotheses regarding other levels. Taking their cue from early studies made by Dale Jackson of the U.S. Geological Survey, Adolphe Nicolas's group at the University of Nantes in France played a fundamental role at this stage. Their studies of the Lanzo peridotite block were an indispensable complement to laboratory experiments showing that this block was plastically deformed by the mechanisms proposed by Goetze and Poirier. Thus, little by little, the viscosity of the mantle began to appear less mysterious. Laboratory study of the deformation of olivine supports—if not macroscopic viscosity—at least the order of magnitude of 10^{19} to 10^{20} poises.

Thus we can see the fundamental ambivalence in the peridotitic mantle. Exposed to brief pressures, such as those of a hammer or an earthquake, it reacts like an elastic and, in the limit, breakable solid. Exposed to high temperatures and enormous pressures for millions of years, it acts like a fluid and is deformed plastically.

These are the fundamental characteristics of this *gigaphysics* of the earth, where time is measured in millions of years, temperatures in thousands of degrees, and pressures in kilobars! No wonder it seemed so improbable to physicists at the turn of the century.

Convection in the Mantle

An indispensable preliminary to all studies of convection, the determination of the properties of fluids made it possible to approach the problem as a physicist would—through experiments and calculations. The physicists of the early twentieth century, especially Lord Rayleigh, Henri Bénard, and Osborne Reynolds, showed that the characteristics of a convective system—such as the geometry of cells, the speed of currents of material, and the stability of structures—can be summed up by a small number of dimensionless numbers that are obtained by combining fundamental physical parameters. For example, the Rayleigh number, which is the relation between the forces that promote motion in a material and those that inhibit it, depends on the size of the system (cubed) and on the thermal conditions at the edges of the system, its density, thermal conductivity, and viscosity. But the fundamental property of the Rayleigh number is that two systems that have different densities and viscosities but the same Rayleigh number behave in exactly the same way. One can immediately understand the geophysicist's interest in this: he can reproduce mantle convection in the laboratory by miniaturizing everything, beginning with time, that famous "geologic time" measured in millions of years that he would like to simulate in a few months.

We will have to abandon our crude experiment of the pot of boiling water to find conditions closer to those of a good simulation. For this we will need boxes made of transparent material so that their insides are visible. The sizes of the boxes are calculated in such a way that they miniaturize the dimensions of the mantle and the duration of the phenomena that take place there at the same time. A low-viscosity liquid, usually oil or glycerine, is put into the box. To induce convection a temperature difference is set up between the top and the bottom of the box, the bottom being hotter. By varying the temperature difference, the dimensions of the box, and the nature of the fluid we can create variations whose consequences for the mobility of the fluid we can observe. These experiments are necessarily only partial simulations, because it is impossible to make them perfectly analogous to actual conditions (or, in technical jargon, to respect the similarity conditions) or to create the geometrical conditions of spherical symmetry. But these simulations nevertheless give us precious information.

This simple example illustrates the role of experimentation in the earth sciences. Actual geologic conditions can never be reproduced in the laboratory, if only because they take place on such a

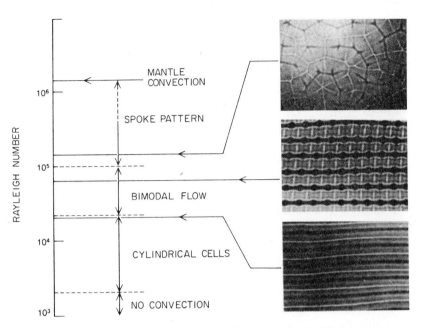

FIGURE 92 The different convective modes that occur in laboratory experiments as the Rayleigh number is increased. Dark lines show descending currents and light lines mark ascending currents. Note that as the Rayleigh number increases the pattern changes from parallel stripes to bimodal stripes and finally to a complex network of convection cells. The first two patterns are stationary, that is, the geometric patterns are stable. The last regime is "turbulent" and the geometric convection pattern continually varies with time.

grand scale. Therefore the earth sciences are not experimental, but observational sciences, in which experimentation can be only a rough guide for theory or observation. Our little desk-top boxes are not the mantle, or even a close approximation of it, but they allow us to study basic phenomena. Also, we have been able to show that when the Rayleigh number increases, the structure of convection changes. When the number is below 3,000, there is no convection. When it is between 3,000 and 30,000, convection takes the form of rising and descending currents in elongated, parallel roll-like convection cells. Above 30,000 motion in the fluid becomes complex, organized in a bimodal structure of a double series of perpendicular rollers. When the Rayleigh number is greater than 200,000, the structure becomes extremely complex, because a series of small cells is superimposed on the large circulation pattern. Moreover, the pattern varies continuously with time, and the convection is unstable. The sequence of events obtained by increasing the Rayleigh number is roughly correct under a wide range of experimental conditions. Thus we can assume that it is equally so for conditions in the terrestrial mantle. Estimates of the Rayleigh number of the mantle, whatever the uncertainties involved in their calculation, show that it is in the non-steady-state range, which means that the mantle has a complex convective structure. The problem of convection in the mantle is not one of the simple problems of fluid mechanics!

Therefore we must abandon experimentation and try calculation, which in this case involves intensive use of the largest computers to simulate the convection conditions of the mantle. Here, as in other endeavors, the computer can play the dual role of an indispensable tool and a sedative for the thought processes. One cannot imagine doing "by hand" the calculations that can be done in a few hours by a modern computer, even if a battalion of humans worked day and night for a lifetime. But the very power of the computer can mesmerize the person who uses it. By infinitely varying the limiting conditions, the user can obtain millions of figures, numbers, and maps from a single program. Drowning in results, he loses his intuitive grasp of the physical phenomena and little by little is carried away from the stated problems, and therefore from their solution. The abundance of figures and curves provides an illusion of scientific rigor but does not encourage the critical scrutiny that is indispensable to the theoretician's task. The computer provides the theoretician with a machine that is the equal of those used by experimentalists in its complexity, price, and high-tech cachet. It demands knowledge and care, but it too easily gives him the feeling that he has covered all the bases. The result has sometimes been a proliferation of complicated studies that cannot be said to have shed any light on the problems at hand.

Some researchers, of course, avoid this trap and are therefore able to domesticate the formidable scientific power of the computer for their own benefit. In the earth sciences a use for the computer that can be called numerical experimentation was gradually developed. This method consists of establishing phenomenological laws by simulating on the computer experiments that are difficult to conduct in a laboratory because they involve, for example, large dimensions and long durations. In the case of convection in the mantle these experiments were extrapolations from the experiments on small models, and they resulted in considerable progress on the problem.

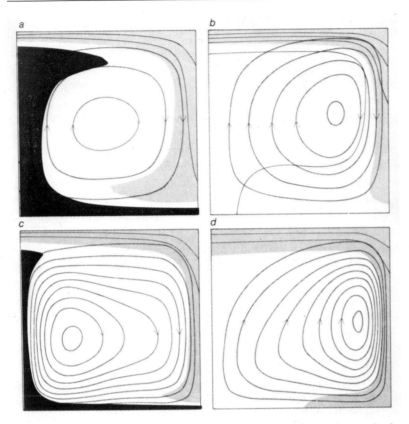

FIGURE 93 Numerical simulations on a computer can be used to study the influences of certain parameters (viscosity, mode of heat generation) on convective flow patterns. When the viscosity is constant and the fluid is heated from below (a), the resulting flow pattern has quasi-symmetrical upwelling and downwelling regions. If the fluid is internally heated (b), there is a strong asymmetry to the flow pattern, with broad upwelling regions and narrow, concentrated zones of downwelling. In (c) the flow pattern changes markedly when the viscosity of the fluid decreases with increasing tempera-ture. In this case upwelling is concentrated within a narrow, hot, low-viscosity channel. The temperature-dependent viscosity effect is less obvious for inter-nally heated fluids (d).

Let us leave technical concerns and get to the essential problem. A solution that is rather popular today was developed by Frank Richter of the University of Chicago and by Dan McKenzie, whom we find once more in the midst of important developments. The terrestrial mantle has two layers, the upper mantle and the lower mantle. Only the upper mantle plays a role in plate movement. It is powered by a complex convection system that features two types of convection. The first, whose typical dimensions are 5,000 to 10,000 kilometers long and 700 kilometers thick, is the one that produces plate movement. The second is smaller, forming cells 800 kilometers long and 700 kilometers thick. The large cells are the product of all the small cells, a bit as the motion of a conveyor belt is only the result of the rollers that propel it. Some of the cell interfaces have edges that are all ascending, the convergence of which produces real chimneys that pump heat from the surface of the lower mantle and create the hot spots that generate volcanic islands. The small cells are much smaller than plates, so the existence of intraplate volcanism is conceivable. The small cells are naturally influenced by the major ridges, with the result that hot spots (upwellings) are found near ridges.

But why invent such a complex mechanism? Why not be content with a single series of large cells taking up the entire thickness of the mantle, rising at the ridges and submerging at the subduction zones? A single layer of cells is unlikely, first of all, for reasons of fluid mechanics: such cells would have extremely elongated shapes—10,000 kilometers long by 3,000 kilometers deep for the Pacific—which would be unusual, because the length and thickness of convection cells always tend to be equal. A more important reason is that organization on several scales is the rule in that region of Rayleigh numbers applicable to the terrestrial mantle. Earthquakes cease to exist below a depth of 700 kilometers, and deep earthquakes are always found in subduction zones. This suggests that the breakable (because cold) plate does not descend more than 700 kilometers, because at that point it reaches the boundary between the upper and lower mantles. Earthquake focal mechanisms indicate that compression takes place at the boundary and, therefore, support this idea. The study of intraplate heat flux shows that the flux is higher than predicted by the model of cooling from the ridges, so it has been suggested that a residual heat flux coming from secondary convective cells should be added to that of the ridges.

Numerical experiments such as these, which led Richter and McKenzie to postulate the existence of two scales of convection, should be added to the picture. But not all geophysicists are convinced. Gerry Schubert of the University of California at Los Angeles, Dick Peltier of Toronto, and Rick O'Connell of Harvard continue to defend the idea of global convection. They make use of additional mechanisms, the simplest of which is the effect of viscosity as it varies with depth. They have used seismological or

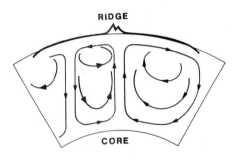

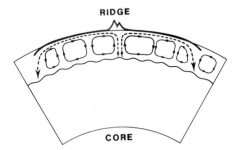

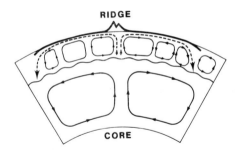

FIGURE 94   Three models of mantle convection. In the top figure the whole mantle convects. In the middle figure, only the upper mantle convects. Finally, in the bottom figure, the upper and lower mantle convect in separate layers; thus there is heat transport but little mass transport between the upper and lower mantle.

gravimetric arguments, but without much success thus far, and it must be said that their phenomenological argument is rather weak. Nevertheless, we have seen other examples of belatedly accepted theories: sometimes truth scorns the arguments of those who search for it! Let us therefore wait and see.

Studies in two fields, however, tend to confirm the two-layer convection model: spatial geodesy and isotope geology.

Isotope Geology and the Structure of the Mantle

Let us go back to the chemical aspects of continental differentiation and extraction from the mantle. Since the continents are very rich in uranium and rubidium compared with the mantle, the extraction of continental material must deplete the mantle of these elements. This impoverishment is reflected in the different isotopic relationships of strontium and lead in the crust and in the mantle. For example, there is a small but significant difference between the isotopic composition of strontium in the basalts of the oceanic *ridges* and the basalts of oceanic *islands*, which led to the conclusion that the ridge basalts and the island basalts come from different reservoirs. Putting this result together with the hot-spot theory, various researchers such as G. Hanson of New York University and M. Tatsumoto of the United States Geological Survey postulated that ridge basalts come from the upper mantle, the part of the mantle that is closely tied to plate movement, whereas island basalts come from the lower mantle.

Quantitative considerations of the isotopic composition of strontium allow us to go even further. The disintegration rate of rubidium 87 must be on the order of a billion years to produce the observed differences in isotopic composition, so the two reservoirs

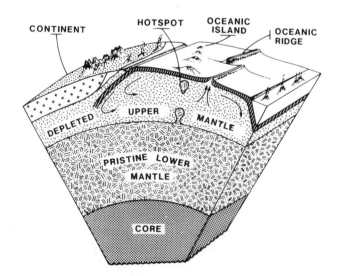

FIGURE 95   Cross-section showing the different chemical reservoirs of the earth. Isotope geochemistry may help us unravel the problem of heat and mass transport among these reservoirs.

must have remained separate for a very long time (on the order of a billion years). The mantle, source of the ridges, being the poorer in strontium 87 (and therefore in rubidium 87), must have formed the continental crust, which is enriched in rubidium, at its own expense. This hypothesis was confirmed by measuring the isotopic compositions of other elements, such as neodymium and lead. Proceeding to a quantitative estimate, Stanley Hart of MIT, my former student Jean-François Minster, and I were able to show in an almost indisputable way that the part of the mantle that had undergone the extraction of the continental crust represented only a third of the volume of the whole mantle. Now one-third of the mantle amounts to an upper layer 700 kilometers thick, which is exactly where earthquakes vanish and where the spinel oxide phase transition takes place (that is, where the peridotite of the mantle undergoes a phase transition from the spinel to the perovskite form, with an accompanying increase in density and S-wave velocity). This is not simply coincidental.

So we come to the idea that the mantle contains at least two layers. The upper layer, 700 kilometers thick, is the source of ridge basalts, and the lower is at least the partial source of island basalts, that is, of basalts originating from hot spots. Hart, Minster, and I also managed to reconcile the ideas of Richter and McKenzie on convection with those of Morgan on hot spots. Will progress in isotopic chemistry make it possible to construct a synthesis acceptable to everyone?

Satellite Geodesy

Contrary to what you might think, the sea surface varies several dozen meters in altitude. Why? This effect is a consequence of the fact that the terrestrial gravitational field is not uniform everywhere: seawater is attracted toward regions where there is an excess of gravity and away from areas that have a gravity deficit. As a result, the sea surface—used as a reference point for terrestrial relief—varies in altitude. In satellite geodesy these differences are measured.

When a current rises from the earth's interior toward the surface it leads to a slight uplift of the surface. The material that is rising is hot and therefore lighter than the surrounding environment. The rise of hot material could be interpreted as a gravity deficit, but in fact the reverse is true, because the difference in density caused by the difference in temperature is not sufficient to compensate for the excess mass created by the uplift. Therefore an ascending

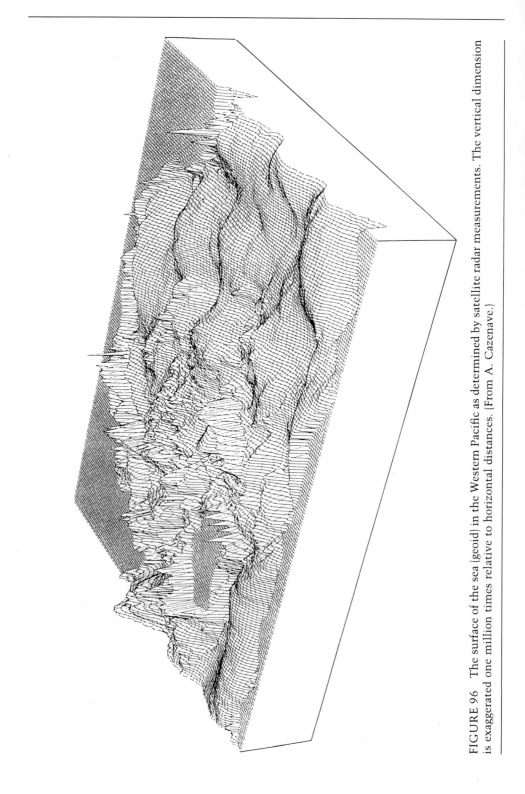

FIGURE 96   The surface of the sea (geoid) in the Western Pacific as determined by satellite radar measurements. The vertical dimension is exaggerated one million times relative to horizontal distances. (From A. Cazenave.)

current corresponds to a positive gravity anomaly, to an intake of water, and a bulge on the sea surface. For descending convection currents the reasoning is naturally the inverse: depressions in the sea surface are seen as indications of descending currents. Therefore the shape of the sea surface reflects both the circulation in the mantle and seafloor morphology; because the latter is a reflection of plate tectonics and therefore of mantle circulation, the two effects tend to reinforce each other.

Submarine volcanoes are excess mass, so the sea surface is higher above them. On the other hand, trenches are deficits of material and therefore of mass. The water is less attracted to them than to other places, so they correspond to depressions. To a first approximation the shape of the sea surface reflects the shape of the bottom.

The map of the height of the sea surface (averaged to cancel out the very local effect of waves) is therefore a clue to circulation in the mantle. It is enough to be able to read these maps to decipher the code; this technique was pioneered at Cornell by Don Turcotte and his students. One studies such maps by successive approximations, first looking only at the large variations, those that extend 20,000 kilometers or more, then at those that are 10,000 to 20,000 kilometers long, and so on—it is like using a zoom lens in photography. The observed image amounts to the superposition not of a series of images but of only two images, a *long-wave-length image* and a *short-wave-length image*. Numerical simulations show that this is exactly what would obtain if there were two layers convecting independently of each other, separated by a border at a depth of 700 kilometers. A detailed study of the short wave lengths brought to light the roll-like cell structure that had been predicted by Richter and McKenzie. But on the other hand it showed that the hot spots are positioned in a way that is largely independent of circulation in the upper mantle, which is contrary to prediction.

Is Convection in the Mantle Well Understood?

In a recent article in *Scientific American** Dan McKenzie of Cambridge stated that convection in the upper mantle is now understood and that we should turn to other geophysical problems. I do not totally share McKenzie's optimism.

To review: The plates move, and their movements vary over time. Convection in the mantle is therefore not of the simple,

* "The Earth's Mantle," *Scientific American*, September 1983, pp. 66–78.

regular type. Hot spots whose origin is deeper than 400 kilometers (the bottom of the asthenosphere) contribute to the complexity of circulation in the mantle. There is no doubt that the cold plates that sink into the mantle are a driving force for surface movement. How, then, do we reconcile all this?

Let us try an analogy that is perhaps daring but not without value. In 1900 the French physicist Henri Bénard demonstrated for the first time the phenomena of convection in the laboratory by heating a thin film of paraffin from the bottom, producing currents that outline the hexagonal forms that are now called Bénard cells. Lord Rayleigh rapidly developed a theory to account for the phenomenon by explaining that since hot fluids are lighter than cold fluids a more or less organized circulation could be established in a fluid heated from the bottom. Today his theory is still the basis of all the theories of convective phenomena.

Now, decades later, we know that the experiment had been badly interpreted. The surface properties of the oil, its surface tension, must be considered along with the thermal phenomenon; these are, in fact, the controlling factors of the Bénard cells' structure. An important theory was built on the foundation of a misunderstood experiment!

The upper mantle, somewhat as in Bénard's experiment, is heated from the bottom, by the radioactivity of the earth and the heat emitted by the earth's core. Also to be taken into account are specific surface phenomena on the cold surface, such as the rigidity of plates or their tendency to sink into the mantle when they are cold. Is it the interaction of these two effects that creates the observed complexity? Whatever it is, there is no doubt that the earth's convection is a very complex physical phenomenon that we must study patiently and observe precisely. Isotope geology and satellite geodesy are two very powerful techniques we can use. Progress in seismology makes us hopeful that it will soon be possible to make a three-dimensional seismic picture of the earth. This will provide essential information for the mapping of the convection cells and increase our power of investigation.

In the meantime, while assessing the considerable progress we have made, we should also measure our ignorance, which is still great. Let us not be like one of the apostles of the tectonics of rigid plates who in 1970 said that geologic problems were virtually all solved and that geology was a completed science. In fact it was only beginning. Earth science has a bright future ahead before it becomes an "exact" science. The era of big discoveries has been opened, not closed.

EPILOGUE

DISCOVERY and persuasion are two goals of science. Discovery is the fortuitous result of the difficult dialogue between man and nature: as Einstein remarked, when one questions Nature, she too often tends to answer "perhaps." Persuasion is the ultimate object of communication with the multitudinous, lively, and abrasive body that is called the scientific community. The researcher is a product of that scientific community; he is imbued with its prejudices, articles of belief, and methods. The goal of his creative imagination, however, is to change the thinking of that community; only after he has convinced the invisible jury of his peers is his discovery integrated into the great body of Knowledge. The example with which I began this book showed that four attempts, all well structured and argued, were necessary to convince the scientific community of the validity of continental drift. In this case it seems that it was more difficult to convince than to discover!

If we are to learn anything from this example, it would be that scientific progress is the result of the interaction between two distinct and complementary entities, the scientific community and scientific thought. Traditional French presentation of science tended to ignore this duality and to claim that ideas, theorems, and theories developed according to a necessary process of reasoning, guided by an inexorable logic, and that the intervention of men and cultures is of only secondary importance. At the other extreme, some thinkers see scientific theories as only an artificial representation of the world, as a purely intellectual product of the human brain, an explanatory analogy of the real world. We know what extreme consequences classical Buddhism and Tibetan Tantrism drew from this view.

In the broadest sense the natural sciences offer the immense advantage of keeping us in touch with the real world, in all its diversity and complexity. They make it possible for us to test scientists' assertions through experiments and observations repeated a hundred times. But when the observed facts and the results of experiments are brought together to form models and theories, human and cultural factors become determining.

Proposing theories and models is the very essence of scientific procedure. Its object is to reformulate the direct apprehension of nature into a synthetic scheme that people can understand. In short, every scientific theory, even the most limited and the most modest, aims to be a universal code enabling us to decipher the messages sent out by Nature. But before this code or syntax can be called a scientific theory, it must first be accepted and assimilated by the *opinion* of the "scientific community." This judgment is difficult to define exactly—and, in any case, it is not up to me to do so—but it turns on the question of what could be called the collective judgment of the scientific world, a sort of conventional wisdom agreed upon by those the community considers its best and best known, its duly dubbed knights. It is formed and externalized in the scientific journals, but even more in the meetings and congresses, by mechanisms that are still difficult to discern but in which personalities and the pedagogical or even polemical qualities of the protagonists play a determining role.

In this example of the theory of continental mobility, we saw Harold Jeffreys lead the entire geological and geophysical community astray by virtue of his "rigor," his eloquence and his prestige as a great scientist. We also saw Fred Vine and J. Tuzo Wilson shepherd the lost sheep back onto the right path by their enthusiasm, their energy, their persistence, and, it must be emphasized, their scientific courage. But of course the role of the scientific community is not simply to judge the validity and admissibility of scientific theories; its principal object is to conceive them, to give birth to them. To use sporting terminology, the whole team referees, and therefore the processes by which discoveries take place are inextricable from the mechanisms that lead to the acceptance or rejection of a theory. I have noted the importance of certain individuals in convincing the really rather stubborn scientific community. At other times the declared opinion of an enthusiastic community pulls along even the most reserved and critical individuals. The implacable opponents of the theory of plate tectonics in 1970, for example, became converts in 1980 and then proceeded to present mobility theory as an eternal certainty, the new converts having rapidly become intransigent devotees. It is the same for processes of discovery and individual actions: small groups and vast communities alternate and combine in a variety of ways for which no precise rule can be defined.

In the discussion of the "sociological" aspects of the procedure followed by earth scientists, the reader will no doubt have noted that certain chapters are very personalized, while others mention few names. By presenting the story in this way I attempted to mirror the reality of events as faithfully as possible. Scientific progress is sometimes the act of an isolated individual, sometimes of isolated groups, sometimes of teams, and sometimes of a numerous scientific community. Wegener was a loner, as was Arthur Holmes; they were both isolated, not in their scientific contacts but in their profound conviction. Progress in paleomagnetism was the work of two groups (at Newcastle and Imperial College), both very much in the minority. The establishment of the chronological scale of reversals of the magnetic field was the product, in the main, of researchers at the University of California at Berkeley. The enormous and decisive progress in the knowledge of the ocean floor was the result of a collective endeavor supported by numerous technical groups and, as a result, by considerable financial means; nevertheless, the names "Scripps" and "Lamont" figure prominently in the history of this work. There, again, two groups of scientists took the decisive steps. Cambridge and Princeton universities were the principal sites in the development of the concepts of seafloor spreading and plate tectonics. At Cambridge at one time or another, under the benevolent direction of Teddy Bullard, were Dan McKenzie, Fred Vine, Drummond Matthews, and John Sclater. At Princeton, where Harry Hess taught, W. Jason Morgan did his thesis and J. Tuzo Wilson and Fred Vine spent a sabbatical year, all during the critical period when the new concepts of seafloor spreading were developing. It seemed as if all the other research centers and universities were simply absent from a moment of decisive scientific progress!

These examples are cited only to show that at certain times during the scientific adventure new initiatives are taken only by a small minority. The size of the minority ranges from one to a dozen. Whether these individuals, groups, or centers are called forerunners, schools, or mafias depends on how well they succeed. But the distinctive characteristic of these pioneers is their isolation. If their research had been submitted to a reasonable consensus of the majority, if they had attempted to achieve a harmonious synthesis accounting for all the opinions, they would never have initiated the decisive new directions that I have described. From experiences of this kind comes—and this is important—the distrust researchers show toward scientific programs defined by consensus, or toward planning research in advance. And because science is often advanced by isolated pioneers, a researcher must believe in his ideas, defend them even when he is in the minority and know how to listen to others' ideas, even when they are the opposite of his own. Of course, all this takes place in a context in which those in the minority are often wrong, in which an intense intellectual competition develops between researchers, and in which the fear of being mistaken haunts everyone oppressively. The courage of a Tuzo Wilson or a Fred Vine no doubt bought ten years for Science; the morally no less respectable courage of Harold Jeffreys cost it thirty!

As soon as the discussion turned to ridges, subduction zones, and tectonics, the names of the scientists became less conspicuous and anecdotes about individuals almost ceased. I recounted the story in this way because progress in these areas was accomplished by a large, active, and enthusiastic community, each member of which contributed something. Once the majority of the scientific community was convinced by mobility theories, it took off in that direction, reaping a harvest of results in a brief period. In the extreme case, the prominence I have given to the work of one individual or another has an arbitrary character that has not escaped my notice; it results merely from an extremely personal sensibility or judgment.

Thus Science progresses, with periods of crisis in which a small group defends its ideas and successfully imposes them, followed by no less active periods in which the whole of the community works toward a common goal. That which was only a sprout becomes a tree of considerable size and diversity. From this powerful and branched trunk sprout offshoots that in turn produce new "trees."

When one looks back after several years of this kind of activity, one realizes that the mass of information, of new discoveries and results, completely submerges the original contributions of the first innovators. When we compare the knowledge we have today of the mechanics of seafloor spreading in all the oceans of the world with Morley and Vine's first efforts at explanation, no doubt their contribution seems small and modest, quantitatively speaking. From this example one could reason that progress is the result of numerous and anonymous groups, that the role of brilliant individuals is only to advance by a few years the dates of discoveries that were inevitable in any case. The history of the idea of continental drift shows that this "collegial" view is contrary to fact. The work of certain individuals was essential, both in repeatedly giving birth to drift theories and in promoting their triumph. On the other hand, we must not think that scientific progress consists of a succession of inspired discoveries, that only the pioneers contribute anything of importance. No isolated individual can replace the richness of creation, development, amplification, and multiplication of an active and enthusiastic scientific community. The view of the earth that we have today integrates the discoveries of thousands of scientists! To debate the relative importance of individuals and large communities in scientific development is pointless. Each plays a role: the individual by proposing new roads, by opposing the ideas of the majority, and by being an "original"; the community by not only acting as a filter (sometimes in the process sacrificing the pioneers), but also by rapidly developing in a standardized and rigorous manner what finally constitutes the great mass of scientific knowledge. At the risk of seeing Science waste away, we must protect this duality, this indispensable antagonism between the individual and scientific society.

Once we have sketched out the relationship between discovering and convincing, which are in my opinion indivisible, we can state that scientific thought does not develop according to the rules that my Cartesian education would like to see it follow.

The most striking aspect of the evolutions I have tried to describe is the disordered nature of the progress of ideas. Blackett's and Runcorn's objective was to establish a curve for the variation of the terrestrial magnetic field in order to understand the origin of the field; in so doing they rediscovered continental mobility! Vine, Matthews, and Morley wanted to draw a magnetic map of the oceans, which led them to support the hypothesis of seafloor spreading. At that time, after the first successes of plate tectonics, it was thought that this theory would rapidly lead to an understanding of the genesis and the history of the continents, the very foundations of geology. Not only did this not happen, but the very tenets of the new theory, first that of plate rigidity and then that of the inactivity of plate centers, were put in doubt. Each step forward, each new movement, is a jump into the unknown. Each theory is an achievement, but also a point of departure, a tool with which to approach new problems. But the solidity of the theories is constantly put to the test, and the tool that appears the sturdiest may break at any instant. In geology as elsewhere, science is an adventure, a never-ending uncertainty. But however chaotic its progress may be, it does not proceed by unforeseeable interventions or new ideas created out of the void. No idea is ever totally new, and in every case one can find the traces of a previous idea. Even the theory of continental drift had its precursors, and in the end there was a certain arbitrariness in my assigning its "origin" to Wegener. I also mentioned Brunhes's and Mercanton's interpretation of reverse magnetism, which was not rediscovered until thirty years later by Cox, Doell, and several others. Morley, Vine, and Matthews's model of marine magnetic anomalies is only an adaptation of the interpretation by Matuyama of the variations of the magnetic field observed in Japanese volcanoes. Geosynclinal troughs and their evolution are found in various versions in modern theories of mountain ranges; the growth of continents from sediments and magmas, in the premises of geology and the eighteenth-century disputes between James Hutton and Abraham Werner. Ideas are modified and transformed, but analysis reveals the old and new elements of each new proposition. Certainly the role of innovation must not be neglected; it is the motivating force of Science, but it would be naive and disastrous to forget a science's historical roots.

Although geology has characteristics that are common to all scientific thought, it has its own specific aspects, the first of which is its character as a historical science. Formerly it was called inorganic natural history, to separate it from the history of living organisms. When, after the pioneering work of nineteenth-century physiologist Claude Bernard, the experimental method was introduced into the natural sciences, the term *Inorganic Natural History*, which had a rather literary connotation, was abandoned and replaced by the *natural sciences*, marking what was thought to be a clear advancement in the classification of the sciences!

Geology, however, is first of all a historical science. Its territory extends not only over millions of square kilometers, but also over 4.5 billion years. Since the phenomena that it studies go far beyond the directly observable or temporal limit, geology's methods are above all historical. Geologists are subject to the uncertainties, the limitations of any attempt to reconstruct the past, sometimes to feelings of vertigo and always to that "myopia" that allows one to grasp much better more recent periods and less and less well the progressively more distant past. Thus mobilist history is limited to the last 200 million years, while the earth is 4.5 billion years old!

Modes of thought among historians of the earth have evolved in a manner parallel to that of historians of human society. As modern history no longer reduces history to a succession of kings and battles, so modern geology has abandoned the dry catalogue of marine transgressions and regressions interrupted by orogenies. As in the "new history," geology today uses quantitative methods and measurements, especially the scientific method based on the formulation of hypotheses grouped into models. Plate tectonics is an example of a science based on this method.

Geology is distinguished from history, in the traditional sense of the term, in that it is a natural science. The messages that have been "written in stone" were written not by man, but by nature. The laws that we seek to discover are not those of human behavior, but of nature. They are more or less complex combinations of the fundamental principles of physics and chemistry. They appear to be extremely complicated and almost inextricable from one another. Natural phenomena contrast with the simple, clearly limited, and well-defined phenomena that the chemist or the physicist can study, experiment upon, and create in the laboratory. Little by little, however, nature is giving up her secrets. Is it possible that these secrets will eventually be explained by "simple" laws, however random? Then again, don't a few simple laws applied to a large number of cases become complex? All human beings have the same number of chromosomes and obey the same laws of heredity, but they are all different. This diversity that biologists correctly stress, and that I have emphasized at various times, is a general characteristic of the natural sciences. It is the result of the accumulation over history of the various simple effects that scientists have so laboriously endeavored to understand.

Before Wegener, geology as a discipline could claim a clearly understood and unchanging temporal framework. The well-defined and fixed continents were bounded by oceans whose movements mysteriously invaded their borders periodically. In the same periodic way, the phases of the geologic cycle inexorably succeeded one another: erosion, sedimentation, folding, granitization, erosion again, and so on. Constantly renewed, the cycles fashioned an earth whose ancient landscapes were thought identical to its recent landscapes. Marine transgressions and regressions, more or less clearly linked to periodic orogenic events, marked this oscillatory stability in continental sedimentation. In this pre-Wegenerian view the concept of time was clearly cyclic, as it was in ancient civilizations, such as Egypt, whose perpetual calendar was calibrated by the annual flooding of the Nile, or the Chou dynasty of China, for which the unchangeable rhythm of the harvest period following the winter season mirrored the alternation of yin and yang, the dualistic representation of the Tao. Starting with Wegener, geologic time became vectoral and unidirectional: it was the irreversible time of history, that which the Greeks, beginning with Herodotus of Halicarnassus, best understood in founding what Pierre Chaunu calls linear history. The continents are mobile, in the

GLOSSARY

Acidic rock   During the early years of geology it was thought that rocks were *salts* resulting from the reaction: acid + base → salt + water. The acid was a hypothetical orthosilicious acid (rich in silica), the bases were limestone (calcium) and magnesia (magnesium). Rocks high in silica (such as granite) were called *acidic* rocks; those rich in calcium and magnesium (such as basalt), *basic* rocks; and those very rich in magnesium (such as peridotite), *ultrabasic* rocks. Today no credence is given to the theory of salts as the origin of rocks, but the nomenclature has remained.

Alpine ranges   A general term used either for almost all the recent mountain ranges (less than 150 million years old) in the Mediterranean area, or for all the earth's ranges between 60 and 20 million years of age.

Alps   The range of mountains that lie in an arc from Provence, France, through Mont Blanc to Austria. The Alps are a product of the collision between Africa and Europe.

Anatexis   The melting of a piece of the continental crust, whether it is sedimentary, metamorphic, or granitic. When melting produces a magma that is granitic in composition, it is called anatexic granite (as opposed to those granites whose origin is deep within the mantle); the very existence of anatexic granite is doubted by many scientists today.

Andesite   A volcanic rock whose chemical composition is intermediate between granite and basalt. It takes its name from the Andean cordillera, where it is particularly abundant.

Asthenosphere   The region of the mantle located 100 to 400 kilometers deep in the earth. Seismic waves are very attenuated in the asthenosphere, and it is generally believed that the movement of the lithospheric plates originates in this soft layer. The rigid lithospheric plates are thought to glide along on the "soft" asthenosphere.

Basalt   The most common volcanic rock, not only on earth but also on the moon (and probably on Mars); it makes up the ocean floor. Basalt is formed by the sudden cooling of a pool of melted silicates on the earth's surface. The minerals in basalt have not had the time to form large crystals and are therefore very small and often not visible to the naked eye. Basalt consists of two essential minerals, plagioclase and pyroxene, and contains other minerals whose presence defines the various types of basalt; among these one of the most common is olivine. Because basalt is "low" in silica (SiO_2) (relative to continental rocks, that is; it is 50 percent silica), it is called *basic*.

Basic rock   *See* Acidic rock.

Benioff-Wadati zone   Location from which, in a given subduction zone, deep earthquakes originate. The existence of this well-defined, sloping geometric plane was discovered by K. Wadati of Japan and confirmed by Hugo Benioff (see Chapters 2, 3, and 4).

Biosphere   The totality of living matter on the earth's surface. It serves as a reservoir for chemical elements such as carbon, sulphur, and nitrogen and affects the phenomena of erosion and sedimentation.

Convection current   A movement of material whose driving force is not local but global and which is part of a more or less cyclic general motion that takes the shape of more or less closed loops.

Diapir   "Bubbles" of rock that pass through the earth's crust. When a light fluid is placed under a heavy fluid (for example, oil under water), the light fluid rises to the surface and produces bubbles. In the same way, when assemblages of rocks are lighter than those lying above them, they tend to rise to the surface by passing through the formations on top of them. This occurs with rock salt (NaCl) formations or volcanic magmas.

Epicenter of an earthquake   The vertical projection on the earth's surface of the locus of an earthquake.

Fault   A fault is a break that offsets two blocks that were initially located face to face. A *normal* fault occurs when the horizontal surface area of the two blocks is greater after the fault than before the fault; this happens when a block is subjected to extension forces. Grabens such as the Rhine and Rhone valleys or the great African lakes are bordered by normal faults, indicating that they were formed by extension movements. A *reverse* fault occurs when the surface area covered by the two blocks decreases after faulting. Such a structure is created when a piece of the terrestrial crust is subjected to compression and reacts by breaking (and not by flexible deformation or folding). A *strike-slip* fault is a fissure that offsets two blocks initially located opposite each other. In strike-slip faults the discontinuity is predominantly vertical, and the movement of the two blocks is therefore horizontal. Strike-slip faults occur when the terrestrial crust is subjected to differential horizontal forces. The forces linked to the spreading of the ocean floor produce large strike-slip faults. The North Pyrenees Fault, the San Andreas Fault, and the New Zealand Fault are all strike-slip faults. A *transform* fault is a fault whose length remains constant and localized. A detailed definition is given in Chapter 3.

Focal mechanism   The orientation and slip vector of the fault plane that slips during an earthquake. The focal mechanism can be determined by examining seismograph records collected at several different azimuthal angles and distances from the earthquake source, or *focus*. This allows seismologists to determine focal mechanisms from earthquakes where we cannot see the earthquake-producing fault at the surface (as is the case with deep earthquakes and earthquakes occurring on the seafloor). Seismologists' ability to determine earthquake focal mechanisms benefitted greatly from the worldwide network of seismometers set up to detect nuclear testing. In turn this new data source provided important proof for the then emerging theory of plate tectonics.

Focus of an earthquake   The place from which an earthquake originates. It can be near the earth's surface or at depth. The vertical projection of the focus on the surface is called the *epicenter*. The focus is also called the *hypocenter*.

Folds   Folds are formed when the terrestrial crust, subjected to pressure, reacts in a flexible way. The formation of folds corresponds to a reduction of surface area. Folds and reverse faults are therefore the characteristic structures of areas subjected to compression, for example, mountain ranges.

Gabbro   A type of basalt that crystallized slowly at depth. It differs from other basalts in that its minerals (plagioclases and pyroxenes) are large and visible to the naked eye.

Geochemistry   The application of chemical methods to the study of the earth. The earth is the scene of vast movements of material and of a series of diverse chemical reactions. One can think of the planet as an immense chemical factory.

Geodynamics   The study of the dynamic aspects of the earth. External geodynamics is the study of the movements of material on the earth's surface, the water cycle, and so on. Internal geodynamics is the study of movements at depth.

Geophysics   The study of the physical phenomena—such as wave propagation or variations in the magnetic field—of the earth.

Gondwana   A continent that during the Permian (250 M.Y.B.P.) consisted of what today is South America, Africa, India, Australia and Antarctica.

Granite   A rock of igneous origin, that is, formed from a pool of melted silicates. The parent magma cools at depth, very slowly, so the minerals in granite have time to crystallize to sizes up to several centimeters across. Granite consists of two major minerals, quartz and potassic feldspar, and contains other minerals, of which the most common are black and white micas. Because the silica content is "high," more than 65 percent, it is called *acidic*. Granite, a coarse-grained rock, is the most abundant rock on the continents. None has been discovered on the moon or in the rocks of extraterrestrial origin called meteorites.

Hydrosphere   The totality of reservoirs of water found on the earth's surface. The hydrosphere includes the oceans, the seas, the lakes, the rivers, the underground waters, and the glaciers, the largest of which are the polar ice caps. The term *cryosphere* is sometimes used to designate the totality of glaciers.

Isostasy   A theory that postulates that the outer layers of the earth, more particularly the crust, float on the part immediately under them, the mantle, like a cork floating in water, following Archimedes' law of hydrostatics.

Isotopes   Atoms that have the same number of external electrons, and therefore the same number of protons in their nucleus, but different numbers of neutrons. Isotopes of the same chemical element have the same chemical properties but different physical properties. Some isotopes are radioactive; others are not (that is, they are stable).

Laurasia   A continent that during the Permian (250 M.Y.B.P.) included what is today North America, Europe, and Asia (minus India).

Limestone   A sedimentary rock composed of calcium carbonate ($CaCO_3$), formed through the accumulation of the shells of microscopic or macroscopic organisms. Limestone rich in magnesium is called dolomite. The weathering of dolomite produces landscapes that look like the ruins of ancient buildings (the North Italian dolomites).

Magmas Liquids formed of melted silicates. Magmas are the source of volcanoes and plutons. When they cool, they produce volcanic or plutonic igneous rocks. Magma is created at depth by the partial melting of the solid materials that constitute the interior of the earth. Their origin can be the crust or the mantle.

Metamorphism A geologic phenomenon during which rock is transformed into another type of rock through an increase in temperature and/or pressure, by solid-state recrystallization without passing through an intermediate stage as a melted silicate. A limestone under great pressure is metamorphosed into a marble; a clay, into schist.

Microtectonics Faults and folds exist on all scales, from a centimeter to dozens or hundreds of kilometers. The study of faults and folds a centimeter to a meter in length is called microtectonics. The goal of this discipline is to understand the mechanisms of folding and faulting and the relations between small structures and those on the scale of a mountain range.

Oceanic ridge High-relief area located in an ocean. In plate-tectonic theory it is the zone in which oceanic lithosphere and therefore terrestrial surface is created (see Chapters 2, 3, and 4).

Orogeny The process of mountain formation, particularly by folding and thrust-faulting. Orogeny occurs during an episode of *orogenesis* and results in the formation of an *orogen* (mountain belt).

P waves Seismic compression waves that generally travel most rapidly and therefore are first to be detected by seismometers, from which comes their name of *primary* waves (see Chapter 2).

Paleogeography The study of ancient geographies, particularly the distribution of seas, lands, and climates. Among the subdivisions of this discipline is paleobiogeography, which is the study of the geographic distribution of extinct organisms that have been preserved in the form of fossils.

Palingenesis The remelting of a rock, from which a new rock is produced having some characteristics inherited from the parent rock. An ancient granite that melts to give birth to a young granite is an example of palingenesis.

Pangaea   The supercontinent formed by Laurasia and Gondwana 250 million years ago. Sometimes the term *Ur continent* is used as a synonym.

Peridotites   Rocks that form the terrestrial mantle, containing more than 50 percent olivine as well as pyroxenes and other accessory minerals such as plagioclase and garnet. The silica content of peridotites is very low: 40 percent; hence they are called *ultrabasic* rocks. The partial melting of peridotites yields a silicate melt of basaltic (basic) composition.

Plutons   Massive rock bodies that result from the crystallization of magma trapped at depth. Granite is an example of a plutonic formation. Its form can be quite varied, but the most common is the laccolith. *Plutonism* is the geologic activity that gives birth to plutons.

Precambrian   The period before the Cambrian, that is, before 550 million years before the present. In classical geology, geologic time was divided into two large and distinct units: the Phanerozoic, from the Cambrian to the present, from which fossils could be traced and used to provide clues to the environment in which they were deposited and to date formations; and the Precambrian, a period without fossils for which only a gross geology could be sketched out. The discovery of radioactivity and its utilization in geology changed that situation. The geology of the Precambrian, which covers seven-eighths of geologic time, is today studied almost as effectively as that of more recent periods.

Radioactivity   Spontaneous nuclear transformation in which one isotope is transformed into another. This transformation, discovered by Antoine-Henri Becquerel in 1896, is independent of all exterior factors and depends upon time alone. Therefore it can be used to record the passage of time and serves effectively as a geologic clock through the use of very long-lived radioactive isotopes (see Chapter 2). Energy is released when radioactive transformations take place. The disintegration of the isotopes potassium 40, uranium 238, and thorium 232 is a source of essential energy for terrestrial phenomena.

Radiochronology   The dating of rocks and geologic formations through the use of radioactive isotopes. The principles of radiochronology are explained in Chapter 2.

Regression A retreat of the sea away from the coasts of the continents. During a regression shore lines are extended.

S waves Seismic shear waves that are called *secondary* waves because they travel more slowly than, and therefore are detected by seismometers later than, primary waves.

Sandstone Sedimentary rock formed by the dehydration and compaction of ancient sand.

SIAL The continental crust, so called because it is rich in silicon and aluminum.

SIMA The mantle, so called because it is rich in silicon and magnesium.

Subduction zone Region in which the oceanic surface sinks into the mantle. This sinking sets off deep earthquakes (*see* Benioff-Wadati zone) and is marked at the surface by the existence of deep oceanic trenches.

Subsidence The sinking of basin floors as sediments are laid down on them. Recorded in the coal-bearing basin in the north of France are wave marks and footprints of terrestrial animals, proofs that these sediments were laid down in shallow water, in which the coming and going of the sea uncovered the beach from time to time. This basin, however, is several thousand meters thick. To explain this dual observation it must be assumed that the basin was sinking during the time that sedimentation was taking place.

Transgression The invasion of the sea onto continental "territory," caused by a change in sea level. Transgressions can be detected in geologic series by the "advance" of shore lines marked by ancient coastal sands.

Ultrabasic rock *See* Acidic rock.

Unconformity A surface of erosion that separates younger strata from older rocks. The presence of an unconformity means that an episode of erosion separated the time of formation of the older rocks from the time of deposition of the younger strata. Unconformities can be used in complicated terrains to date the relative age (younger, older) of the two rock units.

CREDITS

A draft translation of the early chapters of this book was made by Barbara Haskell Kurz, part of which was incorporated in the present version. The author and publisher gratefully acknowledge her contribution.

Figure 3   After Beno Gutenberg, *Physics of the Earth's Interior* (New York: Academic Press, 1959).

Figure 5   From A. Hallam, "Continental Drift and the Fossil Record." Copyright © 1972 by Scientific American, Inc. All rights reserved.

Figure 11   From Xavier Le Pichon, "The Birth of Plate Tectonics," *Lamont-Doherty Geological Observatory of Columbia University, 1985–1986*, 56.

Figure 21   From Seiya Uyeda, *The New View of the Earth*. Copyright © 1978 by W. H. Freeman and Company. Reprinted with permission.

Figure 23   From Seiya Uyeda, *The New View of the Earth*. Copyright © 1978 by W. H. Freeman and Company. Reprinted with permission. (*a*) After A. Cox, B. Dalrymple, and R. Doell, "Reversals of the Earth's Magnetic Field." Copyright © 1967 by Scientific American, Inc. All rights reserved. (*b*) After F. J. Vine, "Spreading of the Ocean Floor: New Evidence," *Science* 154 (1966): 1409. Copyright © 1966 by the AAAS.

Figure 25   From F. J. Vine, "Spreading of the Ocean Floor: New Evidence," *Science* 154 (1966): 1406. Copyright © 1966 by the AAAS.

Figure 27   From J. R. Heirtzler et al., "Marine Magnetic Anomalies, Geomagnetic Field Reversals, and Motions of the Ocean Floor and Continents," *Journal of Geophysical Research* 73 (1968): 2119. Copyright © 1968 by the American Geophysical Union.

Figure 28 (*left*)   From J. R. Heirtzler et al., "Marine Magnetic Anomalies, Geomagnetic Field Reversals, and Motions of the Ocean Floor and Continents," *Journal of Geophysical Research* 73 (1968): 2119. Copyright © 1968 by the American Geophysical Union.

Figure 28 (*right*)   From R. L. Larson and W. C. Pitman, "Worldwide Correlation of Mesozoic Magnetic Anomalies and Its Implications," *GSA Bulletin* 83 (1972): 3645. Reprinted with permission.

Figure 29   Compiled by W. C. Pitman, III, R. L. Larson, and E. M. Herron, 1974. Reprinted with permission.

Figure 30   From Arthur E. Maxwell et al., "Deep Sea Drilling in the South Atlantic," *Science* 168 (1970): 1055. Copyright © 1970 by the AAAS.

Figure 31   From John G. Sclater and Jean Francheteau, "The Implications of

Terrestrial Heat Flow Observations on Current Tectonic and Geochemical Models of the Crust and Upper Mantle of the Earth," *Geophysical Journal of the Royal Astronomical Society* 20 (1970): 509. Reprinted with permission from Blackwell Scientific Publications Limited.

Figure 32   From M. Barazangi and J. Dorman, "World Seismicity Map Compiled from ESSA Coast and Geodetic Survey Epicenter Data, 1961–1967," *Seismological Society of America Bulletin* 59 (1969): 369. Reprinted with permission from the Seismological Society of America.

Figure 33   From Robert S. Dietz and John C. Holden, "The Breakup of Pangaea." Copyright © 1970 by Scientific American, Inc. All rights reserved.

Figure 41   From M. Barazangi and J. Dorman, "World Seismicity Map Compiled from ESSA Coast and Geodetic Survey Epicenter Data, 1961–1967," *Seismological Society of America Bulletin* 59 (1969): 369. Reprinted with permission from the Seismological Society of America.

Figure 42   From Bryan Isacks, Jack Oliver, and Lynn R. Sykes, "Seismology and the New Global Tectonics," *Journal of Geophysical Research* 73 (1968): 5855. Copyright © 1968 by the American Geophysical Union.

Figure 47   From Donald L. Turcotte and Gerald Schubert, *Geodynamics: Applications of Continuum Physics to Geological Problems* (New York: John Wiley and Sons, 1982). Copyright © 1982 by John Wiley and Sons. Reprinted by permission of John Wiley and Sons.

Figure 48   From J. Besse, "Cinématique des plaques et dérive des pôles magnétiques." Thèse d'Etat, Université Paris VII, Paris, France, 1986.

Figure 63   From Seiya Uyeda, *The New View of the Earth*. Copyright © 1978 by W. H. Freeman and Company. Reprinted with permission. (*b*) After B. Isacks and P. Molnar, "Mantle Earthquake Mechanisms and the Sinking of the Lithosphere," *Nature* 223 (1969): 1121. Reprinted by permission from Nature. Copyright © 1969 by Macmillan Magazines Limited.

Figure 64   From Seiya Uyeda and Hiroo Kanamori, "Back-Arc Opening and the Mode of Subduction," *Journal of Geophysical Research* 84 (1979): 1049. Copyright © 1979 by the American Geophysical Union.

Figure 65A   From Seiya Uyeda, *The New View of the Earth*. Copyright © 1978 by W. H. Freeman and Company. Reprinted with permission. After Dan E. Karig, "Evolution of Arc Systems in the Western Pacific." Reproduced, with permission, from the *Annual Review of Earth and Planetary Sciences*, vol. 2. Copyright © 1974 by Annual Reviews Inc.

Figure 65B   From Seiya Uyeda, *The New View of the Earth*. Copyright © 1978 by W. H. Freeman and Company. Reprinted with permission.

Figure 67   Reproduced from *Evolution of Sedimentary Rocks*, by Robert M. Garrels and Fred T. Mackenzie, by permission of W. W. Norton and Company, Inc. Copyright © 1971 by W. W. Norton and Company, Inc.

Figure 70   From David T. Griggs, "A Theory of Mountain Building," *American Journal of Science* 237 (1939): 611.

Figure 71   From J. F. Dewey, "Continental Margins: A Model for Conversions of Atlantic Type to Andean Type," *Earth and Planetary Science Letters* 5 (1969): 189.

Figure 72 (*top*)   From Peter Molnar and Paul Tapponnier, "Active Tectonics of

Tibet," *Journal of Geophysical Research* 82 (1978): 5361. Copyright © 1978 by the American Geophysical Union.

Figure 72 (*bottom*) From Peter Molnar and Paul Tapponnier, "The Collision between India and Eurasia." Copyright © 1977 by Scientific American, Inc. All rights reserved.

Figure 73 (*top*) From David E. James, "The Evolution of the Andes." Copyright © 1973 by Scientific American, Inc. All rights reserved.

Figure 74 (*top*) From B. Clark Burchfiel, "The Continental Crust." Copyright © 1983 by Scientific American, Inc. All rights reserved.

Figure 77 From Stephen Moorbath, "The Oldest Rocks and the Growth of Continents." Copyright © 1977 by Scientific American, Inc. All rights reserved.

Figure 79 Reproduced from *Evolution of Sedimentary Rocks,* by Robert M. Garrels and Fred T. Mackenzie, by permission of W. W. Norton and Company, Inc. Copyright © 1971 by W. W. Norton and Company, Inc.

Figure 88 (*top*) From Kevin C. Burke and J. Tuzo Wilson, "Hot Spots on the Earth's Surface." Copyright © 1976 by Scientific American, Inc. All rights reserved.

Figure 88 (*bottom*) From G. Brent Dalrymple, Eli A. Silver, and Everett D. Jackson, "Origin of the Hawaiian Islands," in *Earth's History, Structure, and Materials: Readings from American Scientist,* ed. Brian J. Skinner, 112.

Figure 89 From Seiya Uyeda, *The New View of the Earth.* Copyright © 1978 by W. H. Freeman and Company. Reprinted with permission. After J. Morgan, "Deep Mantle Convection Plumes and Plate Motions," *AAPG Bulletin* 56 (1972): 203.

Figure 91 From D. Forsyth and S. Uyeda, "On the Relative Importance of the Driving Forces of Plate Motion," *Geophysical Journal* 43 (1975): 163. Copyright © 1975 by Gordon and Breach Science Publishers Inc. Reprinted with permission of the authors and publishers.

Figures 92 and 93 From Dan McKenzie and Frank Richter, "Convection Currents in the Earth's Mantle." Copyright © 1976 by Scientific American, Inc. All rights reserved.

Figure 96 Reprinted with permission from Anny Cazenave.

Plates 2 and 4 From John G. Sclater, Barry Parsons, and Claude Jaupart, "Oceans and Continents: Similarities and Differences in the Mechanisms of Heat Loss," *Journal of Geophysical Research* 86 (1981): 11,535. Copyright © 1981 by the American Geophysical Union.

Plate 3 Map of the world ocean floor by Bruce C. Heezen and Marie Tharp. Copyright © 1977 by Marie Tharp. Available from Marie Tharp, One Washington Avenue, South Nyack, New York 10960.

Plate 5 Map appears courtesy of Paul Hoffman, Geological Survey of Canada, Ottawa, Ontario.

INDEX

Intercontinental bridges, 7–8
International Phase of Ocean Drilling (IPOD), 133, 207
Iran, 111; mountains, 175, 207
Irving, Ted, 45–46, 48, 208
Isacks, Bryan, 99, 105–108, 121, 155, 158, 178
Island arcs, 5, 14, 30, 81, 155, 163–164, 166, 206, 209
Isostasy, 1–4, 8, 15, 230
Isotope geology, 216, 237, 242; as chronometer, 34–37, 221; and continental growth, 201–205; and structure of mantle, 238–239. *See also* Geologic clock

Jackson, Dale, 231
Jacob, François, 198
Japan(ese), 28, 104, 155, 157; island arc, 5, 209; mountain ranges, 13; Sea of, 14; volcanism, 109, 153–154; subduction zones, 153–154
Jaramillo event, 54
Jaupart, Claude, 206
Jeffreys, Harold, 17–18, 25, 27, 52, 89, 171, 217, 244, 246
JOIDES (Joint Oceanographic Institutions for Deep Earth Sampling), 75, 133
Joly, John, 16, 220
Jordan Fault, 152, 211
Juan de Fuca Ridge, 71, 105

Karig, Dan, 164
Kay, Marshall, 172, 174
Kelvin, Lord, 37
Kermadec Islands, 99, 155
Kilauea, 221
Kilimanjaro, 87
Koenigsberger, Johann G., 45
Kossmat, Franz, 12, 120, 171
Krafla volcano, 146
Krakatoa, 158
Kuno, Hisashi, 154
Kuno's conjecture, 154, 159
Kurile island arc, 5, 104, 155, 157, 160, 206
Kurile trench, 78
Kushiro, I., 120

Lamont Geological Observatory, 28, 31–32, 54, 69, 71, 72, 74, 86, 90, 99, 108, 128, 138, 155, 178, 213, 245

Lanzo peridotite block, 231
Laramide range, 198
Larochelle, Andre, 69
Lau basin, 164
Laurasia, 5, 8, 13, 111, 134
Lava flows, 46, 54
Lehmann, Inge, 27
Le Pichon, Xavier, 69, 98–99, 108, 113–114, 120, 124
Lesser Antilles island arc, 81, 155
Liboutry, Louis, 228
Limagne graben, 208
Lister, Cliff, 149
Lithosphere, 61, 104, 106, 138–140, 142, 151, 183; lithospheric plates, 111, 144, 155, 164, 217
Lugeon, Maurice, 12
Lyell, Charles, 11, 200

Magma, 5, 30, 38, 168, 204–205, 222; birth of, 39; volcanism and, 109; heat transfer by, 149; andesitic, 163; basaltic, 163
Magma chamber, 144
Magmatic phenomena, 109, 200
Magnetic anomalies, 42, 44, 74, 82, 133–134, 164, 222; in Japanese lava, 53, 64; marine, maps of, 30, 63–70, 86, 119, 123, 127; calculation of, 65; pattern-displacement of, 70; symmetrical, 71; and measurement of spreading rates, 72; chronology of, 99; and plate tectonics, 106. *See also* Seafloor spreading
Magnetic dating, 72, 74, 82
Magnetic equator, 66
Magnetic field, terrestrial, 41–46; inversions of, 52–57, 64, 71, 248
Magnetite, 45
Magnetometer, 63
Mantle, 18, 25, 29, 61, 75, 78, 104, 138, 148; thickness of, 27; and basalt, 40; recycling in, 60; and plate movement, 106–107; melting of, 109, 142, 159; composition of, 139, 189, 190, 192, 231; and chemistry of seawater, 148–149; plunging of cold plate into, 155, 157; large ions from, 196; properties of, 229–231; two layers, 239–240; circulation, 241. *See also* Convection
Mantle-core boundary, 27
Marginal basins, 163–166

Ocean-continent ratio, 201
Ocean floor: topography of, 32–33; subsurface of, 33; as conveyor belt, 60–61, 65, 75–76, 133, 137, 160; magnetic fields of, 63; transform faults, 70; magnetic dating of, 72, 74, 82; composition of, 123; and history of oceans, 123–131, 133–137; morphology of, 131; structure of, 138. *See also* Oceanic crust; Seafloor spreading
Oceanic basins, 155, 170
Oceanic crust, 30, 71, 78, 80, 189; composition of, 33; structure of, 138–140, 142; formation of, 144; hot-water circulation, 148; recycling through, 149; creation of, 158. *See also* Crust; Seafloor spreading
Oceanic ridges, 33, 76, 88; zones, 94; earthquakes along, 104; volcanic, 138; emerged, in Iceland, 144; surface creation, 164; isotopic composition of basalts of, 238. *See also* Ridges
Oceanic trenches, 94, 104–105, 161–162; gravitational anomalies along, 154. *See also* Subduction zones; Trenches
Oceanography, 15, 63; geological, 30–34; seafloor spreading, 117
Ocean stratigraphy, 75, 133–134, 137
O'Connell, Rick, 236
O'Hara, Mike, 40
Okhotsk Sea, 14
Oldham, Richard D., 24–25
Olduvai event, 54
Oliver, Jack, 99, 105–108, 121, 178
Olivine, 37, 189, 192–193, 230–231
O'Nions, Keith, 203, 205
Opdyke, Neil, 54–56, 72
Ophiolites, 63, 139–140, 144, 175, 183, 185, 186; ophiolitic massif, 147; ophiolitic belt, 209, 211
Orogeny, 140, 167, 170, 181, 198, 249; orogenic belts, 193–196, 200, 203, 206, 209, 211; function of, 196
Oxburgh, Ron, 144, 220

Pacific Ocean, 95, 209; trenches in, 33, 81, 104; magnetic inversions in, 54, 134; seismic belt in, 81; fracture zones in, 97; subduction zones in, 104, 134; history of, 134–137; mountain ranges surrounding, 175
Pacific plate, 119, 222, 224
Pacific ridge. *See* East Pacific Rise
Paleogeography, 22
Paleomagnetism, 46–54, 57, 88, 208–209, 245; chronology and, 49, 72
Paleontology, 7–8, 15, 17, 57; stratigraphic, 8; micro-, 56. *See also* Fossils
Palingenesis, 40, 198
Pangaea, 4, 8, 18, 49, 86, 116, 124, 194, 208
Papua New Guinea, 140; mountains, 175, 183–184
Parker, Robert, 95–96, 99, 106, 120
Patriat, Philippe, 124
Patterson, Clair, 202–203
Peltier, Dick, 230, 236
Peridotite, 37, 59, 139–140, 142, 144, 189, 193, 217, 230–231
Peru, 104, 109, 155, 183, 209
Petroleum, 21–22, 134, 137, 212, 214, 215
Petrology, 37, 120, 160, 217; experimental, 39, 41; volcanic, 154
Philippines, 109
Pillow lavas, 139, 144–145, 147
Pitman, Walter, 69, 72, 124, 128, 137, 213
Plagioclase, 37
Plate boundaries, 129, 138–166, 175; Q factor at, 106; and geologic activity, 111, 138, 175; mapping of, 113
Plate creation, 164
Plate geology, 13, 121, 138
Plate movement, 216–242; and convection, 216–221; mechanisms of, 226–229; and gravity, 227–229
Plate tectonics, 30, 61, 92, 123–128, 172, 247; and spherical geometry, 93, 99; rigidity of plates, 93, 99, 106; pole of rotation, 94, 99; and earthquakes, 99–109; and magnetic anomalies, 106; and plate boundaries, 106, 111, 113; and volcanism, 109–110; laws of, 111, 113; reactions against, 117–121; and global geology, 122; and history of oceans, 137; and tectonics, 174; intraplate movements, 179; and mantle convection, 220, 241